This document has been approved for public release, with redactions, by the Public Release Review Team at the National Security Agency. The case number is PP-18-0649.

Printed in the United States of America
First Printing, 2019

Foreword

I've written this memoir for two reasons. First, because my own father passed away before I got to hear many stories about his own military service during World War II. I think that he was reluctant to tell many stories when I was younger, as the most interesting parts of his service involved classified operations with the OSS in Burma. His history probably included very stressful situations, hardship and poor rations, as well as some combat. My story includes no combat and only occasional poor rations but does include a lot of stressful situations.

The second reason that I'm writing this is to document the efforts of the people I worked with in the Naval Security Group and aboard the USNS Wheeling and the Claud Jones class destroyer escorts. I worked with these people on projects that were classified at the time. Some of the details of the projects have been declassified and appear in the public record. Other aspects of the projects, in particular, recent Pony Express operations, remain classified.

At the end of the document, I have added appendices with more technical information from public domain unclassified or declassified sources.

TABLE OF CONTENTS

CHAPTER 1	THE BEGINNING	1
CHAPTER 2	BOOT CAMP	3
CHAPTER 3	DLIWC	9
CHAPTER 4	RADIOTELEPHONE OPERATOR TRAINING	19
CHAPTER 5	TRAFFIC ANALYSIS AND METADATA	24
CHAPTER 6	SIDI YAHIA MOROCCO	26
CHAPTER 7	EUROPEAN LEAVE	35
CHAPTER 8	NAVOCS NEWPORT R.I.	40
CHAPTER 9	PENSACOLA TO OAKLAND	51
CHAPTER 10	NAVCOMMSTA HONOLULU	54
CHAPTER 11	RPIO AND TGU PEARL HARBOR	64
CHAPTER 12	UNCLASSIFIED LIFE	67
CHAPTER 13	USNS WHEELING AND HULA HOOP	76
CHAPTER 14	WHEN WAS OUR LAST HAIRCUT, SAILOR?	106
CHAPTER 15	HULA HOOP POST-MISSION BRIEFINGS	108
CHAPTER 16	PONY EXPRESS	110
CHAPTER 17	SPOOFING THE RORSAT	135
CHAPTER 18	NEW ORDERS AND READJUSTMENT	138
CHAPTER 19	AFTER SECGRU	141
CHAPTER 20	JOIN THE NAVY, SEE THE WORLD	145
APPENDIX A.	WULLENWEBERS, SIGINT AND HFDF	147
APPENDIX B.	DECLASSIFIED HULA HOOP DOCUMENTS	163
APPENDIX C.	THE OUTSIDER VIEW OF NUCLEAR TESTING	172
APPENDIX D.	EXCERPT FROM FITNESS REPORT	174

TABLE OF FIGURES

Figure 1 Boot Camp Petty Officers. 7
Figure 2 Receiving my Academic Award 8
Figure 3 View from DLI dorm fire escape. 16
Figure 4 DLI Graduation Certificate 17
Figure 5 CT Rating Badge. 18
Figure 6 R-390 HF Receiver front panel 22
Figure 7 Radiotelephone Operator Certificate 23
Figure 8 Aerial view of main station at Sidi Yahia 34
Figure 9 Gelato in Rome 39
Figure 10 My First DD-214 49
Figure 11 OCS Commissioning Portrait 50
Figure 12 High point of Cross-Country drive 53
Figure 13 The places I worked on Oahu 62
Figure 14 FRD-10 Antenna at NavCommSta Honolulu 63
Figure 15 Rehearsal photo from After the Fall 74
Figure 16 Rehearsal photo from Born Yesterday 75
Figure 17 A daily menu from the Wheeling 99
Figure 18 Shellback card from first equator crossing 100
Figure 19 Posing near the pelorus on the bridge of the Wheeling 101
Figure 20 Wheeling at the dock in Pago Pago 102
Figure 21 Helicopter from USNS Corpus Christi Bay delivering mail. 103
Figure 22 French Minesweeper Dunkerquoise 104
Figure 23 The Helicopter repair ship USNS Corpus Christi Bay. 105
Figure 24 SECGRU Division on the USNS Wheeling 107
Figure 25 The USS McMorris in 1972 128
Figure 26 Layout of the Telemetry room. 129
Figure 27 Telemetry intercept position from the 1970s 130
Figure 28 MUC Citation for Operation IVY GREEN 131
Figure 29 Soviet KA-25 helicopter near the USS Claud Jones 132
Figure 30 USS McMorris approaching the Claud Jones 133
Figure 31 SMRIS Chazma 134
Figure 32 My travels in the Navy 146

Ranks, Ratings, Acronyms, and code names

Here are a few things you need to know to make sense of the rest of this document. As is the case in most large organization, the Naval Security Group was replete with organizational titles and acronyms.

NSG or SECGRU Naval Security Group. This was the Navy organization charged with Signal Intelligence (SIGINT) and Communications Security (COMSEC) operations. It was formed in about 1947. The name was changed to Information Warfare Operations in the1990s.

Navy Ranks (Communications Technician in parentheses)

RATES are enlisted specialties, of which Communications Technician is one. RANK is the pay grade within a rate or officer specialty

E1	Seaman Recruit	Normal rank during boot camp
E2	Seaman Apprentice (CTISA)	Normal rank after boot camp
E3	Seaman (CTISN)	6 months as E-2 and passing exam
E4	Third Class Petty Officer (CTI3)	6 months as E-3 and passing exam
E5	Second Class Petty Officer (CTI2)	1 year as E-4, passing exam, selection
E6	Petty Officer First Class (CTI1)	Several years, tough exams, selection
E7	Chief Petty Officer (CTIC)	Lots of years, tough selection, etc.
E8	Senior Chief (CTICS)	
E9	Master Chief (CTICM)	Highest enlisted rank. Less than 1% of enlisted
01	Ensign	Lowest officer rank
02	Lieutenant Junior Grade (LTJG)	Automatic promotion for Ensign after 18 months.
03	Lieutenant	What a LTJG becomes after 3 years, if there are no issues.
04	Lt. Commander	Mid-level managers. Small ship commanding officers.
05	Commander	Large ship and shore station department heads

General Navy Terms

Chief	Petty officer E7 and above. Chiefs are the experienced enlisted leaders of all Navy organizations. Junior officers are expected to learn from and consult with their chiefs before making important decisions.
CINCPACFLT	Commander in Chief, Pacific Fleet
Department	Group of divisions with similar tasks. Aboard ship there are Weapons, Deck, Operations and Supply Departments. At shore stations there are Operations, Supply NSG, etc. NSG department head is usually a commander or captain.
Division	Administrative group performing a specific task. Usually about 30 sailors, one or two chiefs, and a division officer. Division officer is usually LTJG or Lt.
NAVOCS	Naval Officer Candidate School. Trains college graduates to become Naval officers.
OC	Officer Candidate. Student under instruction to become a commissioned officer.

Communications Technician specialties ca. 1970

CTA	Administrative Branch Handles administration and paperwork. Keeps records and handles office chores.
CTI	Interpretive Branch Linguists. Collect voice data and do translations.
CTM	Maintenance Branch Maintains collection and cryptographic equipment

CTR	Collection Branch Collects Morse code and other non-voice data
CTO	Operations Branch Handles sensitive communications for NSG
CTT	Technical Branch Collects non-voice data such as teletype and telemetry

SIGINT Code Names and Acronyms

NSA	Department of Defense agency tasked with signal intelligence and communications security (US codes and secure communications).
DIRNSA	Director, National Security Agency. The military officer in charge of SIGINT.
COMNAVSECGRU	Commander, Naval Security Group. The head of Naval SIGINT.
DIRSUP	Direct Support. SIGINT collection designed to support ship and fleet commanders in accomplishing their missions.
National Tasking	SIGINT collection tasks to support national, as opposed to local command objectives.
HFDF	High Frequency Direction Finding. Location of signal emitters with multiple specialized antenna and receiver systems.
FRD-10	Large circular antenna array for HFDF. Also called Wullenweber, after the inventor, or Elephant cage. You could see the Wahiawa FRD-10 from the Dole Pineapple display on the way to the North Coast of Oahu.
PELAGIC	Code name for Navy HFDF program primarily used to track Soviet submarines.
BYEMAN	SIGINT from satellites. Later used only in contractor documents
TALENT/KEYHOLE	Satellite based imagery and SIGINT.
DELTA	
GAMMA	Particularly sensitive SIGINT operations

PONY EXPRESS	[redacted]
POINTED FOX	The destroyer-escort program for Pony Express.
HULA HOOP BURNING LIGHT	Programs to collect data from French atmospheric nuclear tests.

Chapter 1 The Beginning

I moved to Corvallis in the Fall of 1968 to attend graduate school in Oceanography at Oregon State University. I had some savings from a summer job, but probably not enough to pay food, rent, and tuition through the fall quarter. I found a short-term part-time job doing housecleaning for a faculty member. Sometime in October, I got a partial research assistant position building equipment for a professor in Atmospheric Science. At that time, I thought I just might last the quarter.

In mid-December, I received a Notice of Induction (draft notice) and was to report for a pre-induction physical exam at the end of December. In 1968, student deferments expired after four years of college, and I had been classified I-A (Available for induction) after graduation. Earlier in the summer, I had applied for Navy Officer Candidate's School, but was not accepted.

After finishing final exams, I told my major professor that I intended to return to OSU after my military service. The university had a policy that students whose studies were interrupted by military service would be automatically re-admitted at the end of their service.

I was living in an upstairs apartment in the house of an elderly widow about a mile and a half from campus. I packed up my apartment and carried my stuff down to my car. Everything I had fit easily into the back of my 1962 Chevrolet Corvair. I told my landlady about the draft notice and that I would not be back after the end of the year and she should send my deposit to my parents. With all my possessions in the back of my Corvair, I drove to my parent's home in Arcata and arranged for my pre-induction physical examination. My mother was researching alternatives to the expected two-years of service in either the Army or Marine Corps.

After passing my physical exam, I had several options:

1. Wait for induction into either the Army or Marine Corps and serve a two-year enlistment which would very likely involve a tour in Vietnam.
2. Enlist in the military for a period of four years, choosing the branch of service myself.
3. Do the draft-dodger thing and move to Canada.

After considering the research done by my mother, I chose option two. My father would have preferred that I enlist in the Air Force—he had been in the Air Force Reserve since the end of WWII. I picked the Navy because of my interest in oceanography and my time in the Sea Explorers in high school. There was a great benefit in the Sea Explorer time in that the manual for the ssea couts is practically identical to the recruit manual that the Navy uses. As a result, I already knew most of the 'book learning' that I would get in recruit training.

Because I had a college degree and had enlisted for four years, I was in a program that would automatically earn me Seaman (E-3) rank after boot camp. This meant a significant pay raise and greater advancement possibilities during my enlistment.

Chapter 2 Boot Camp

On the 6th of January 1969, I caught a bus, with about thirty other inductees, at the Humboldt County Courthouse. About 20 hours later I was deposited on a wet stretch of pavement at the Navy Recruit Training Center in San Diego. The next twelve weeks of training, called boot camp, were designed to give me basic military and naval training and prepare me for either an advanced training position or service in the fleet.

It was a cold and rainy winter in San Diego. In February the average monthly rainfall in San Diego, CA is 1.56 inches with rain usually falling on 7 days. In February 1969, there was quite a bit more rain—over 16 days. We marched through a lot of puddles!

The first few days of boot camp included a lot of standing in line: waiting for our slot in the mess hall, waiting to get lots of shots, and waiting for uniforms to be issued. The last of these waits included waiting for about 30 older sailors to get a special uniform issue. They were part of the crew of the USS Pueblo, a signal intelligence ship that had been captured by the North Koreans. They had just been released from the Balboa Naval Hospital after returning from 16 months captivity in North Korea. I remember thinking that I didn't want a job with that kind of risks. Little did I know that I would spend a lot of time over the next 5 years on shipboard signal intelligence missions.

Since I was one of only two college graduates in my recruit company, I was assigned two additional duties: Mail Petty Officer, and Educational Assistant. The Mail PO job had a good fringe benefit: instead of waiting in the mess hall line with the company at lunch, I used a separate line for single recruits—which moved much more quickly. That gave me time to get to the Post Office and pick up the company mail, which I distributed after lunch. The Educational Assistant job also had a fringe benefit. Along with the other college grad, I was bunked in a two-person room separate from the rest of the barracks which housed the other 40 recruits. This room was also used to do some after-taps additional training for recruits struggling with the academic material. Almost all the struggling recruits were Filipinos who had entered the Navy as a quick path to US residence and citizenship. Some of them had only minimal English skills. With the help of some of the better-educated Filipinos, I helped them improve their language skills and tutored them in the Navy organizational material and vocabulary. These guys were all volunteers and were highly motivated and I enjoyed working with them. At that time, the Navy only allowed these recruits into service jobs, such as cook, steward, and ship's service, barbers, storekeepers, etc.

The other college grad with whom I shared the room only lasted a few weeks. He had used a lot of LSD while in college and started having flashbacks and mental issues. He got a medical discharge after about three weeks in Boot Camp.

Boot camp wasn't a terrible experience for me. The bad weather limited the amount of time we could spend outside marching and on the obstacle course. We were told we were very lucky, as the Marine recruits on the other side of the fence made no allowance for bad weather and their boot camp was much more physically demanding. We did have to spend some time each day outside at the laundry tables washing our own clothes for the first four weeks. After that we got laundry service. Others in the company had to spend a week of mess hall duty—mostly washing dishes. I was exempt from that as Mail and Educational PO. I had learned most of the academic material on ship types, naval organization and rank structure while in Sea Explorers, so I tutored instead of memorizing.

One of the highlights of the first few weeks was the day we spent taking the military aptitude tests. There were two tests, one on general intelligence and aptitude and another on mathematical and logic aptitude. The tests were known as the GCT/ARI, for General Comprehension Test and Arithmetic test. I did very well on the tests---I think I only missed one question on each of the tests. This qualified me for any of the technical ratings in the Navy. After those results were available, I also took a language aptitude test, which I also passed with flying colors, and an electronics aptitude test. The language test results surprised me a bit as I had never thought I had a great aptitude for languages. I had struggled with the required one semester of German in college and thought that the high school German classes were a lot of work. I got good grades in high school German, but never really impressed the teacher. As a result, it was a bit of a surprise to all when I got the best score in the class and won a prize in a national test for students of German. I did well enough on the military language test that I was called in for an interview and asked if I would like to apply for the Communications Technician rating. I was told that most of the job description was classified, but that it would include six months or more of language training and a large percentage of shore duty assignments. I agreed and was asked to fill out a background investigation form to start the security clearance process. The form required references and family history, including all the places I had ever lived. I had to write a letter to my Mom and get the history items, as I had lived in about 10 different places before our family moved to Arcata in 1957.

Another highlight of Boot Camp was the firefighting training. All naval recruits get about 20 hours of firefighting theory and practice. The final exercise is controlling a fire in a simulated ship compartment. Over the course of a morning, the instructors ignite about a half-gallon of oil and gasoline mixture in the compartment, then send the recruits in to douse the mixture. The firefighting training was visible all over the camp as it resulted in a large cloud of black smoke for each simulated fire. It's quite exciting at the time, but the aftermath involves a lot of sooty snot and blackened fatigue uniforms. A quick training session with gas masks and tear gas is also part of the firefighting school. The final part of the exercise takes place in a compartment with a tear gas bomb. You are fine until you must take off your mask and recite your name and serial number before exiting the room. Lots of coughing and spitting ensues!

We also got to spend a couple of days on the firing range which was at a nearby Marine Corps base. We fired WWII vintage M-1 rifles at targets 200 yards away. When we weren't shooting, we were crouched in the pits behind the targets. We would pull the targets down so a petty officer could record the shooter's scores, then we patched the bullet holes for the next round. You could hear the bullets going over your head. One inattentive recruit reached up for a target at the wrong time and was tackled by the range petty officer. I thought our company commander had covered the range of Naval Invective, but we learned some new phrases that day. The range petty officers also demonstrated fully automatic fire with a Thompson submachine gun, and a few lucky recruits got to fire one. I wasn't one of the lucky ones.

At the end of boot camp, we had a graduation parade and were granted a day of liberty—our first free time off base in 12 weeks. As the company academic honor man, I got an extra day of liberty before shipping out to my next duty station. We were warned about some of the shadier areas of San Diego---the Gaslight district in particular. (This was well before the cleanup and gentrification of that area.) Most of the recruits were under 21 and could not legally purchase alcohol. However, many bars would serve beer to anyone in uniform. As I was 22, I could drink in classier places, but limited myself to a few beers and spent most of my time just walking around enjoying the freedom to go where I pleased.

On our last day in San Diego, we were given travel orders to our next duty station and transportation vouchers to get us home on leave, then to our next station. We were allowed a week of leave before reporting to the next station. I took a bus to Arcata, spent a week with my parents, then drove my car south to the Defense Language Institute in Monterey California, where I was to study French for six months.

COMPANY COMMANDER B. R. BOLEN, EMCS AND PETTY OFFICERS - 12 MARCH 1969.

Figure 1 Boot Camp Petty Officers.
I am at right in the top row.

Figure 2 Receiving my Academic Award

Chapter 3   DLIWC

The Defense Language Institute West Coast is located on the grounds of the Presidio of Monterey in the hills above downtown Monterey. I arrived there in May of 1969 and was assigned a double room in the Company C barracks. The barracks was much like a college dormitory, with double rooms having two beds, two desks and two lockers. These were quite luxurious quarters for junior enlisted personnel. My room was on the north side of the third floor, and I had a nice view out over Monterey Bay.

DLI Language classes range from 24 to about 80 weeks in length. The more difficult language, such as Chinese, Russian, Arabic, etc. have longer classes because they require instruction in new writing systems or alphabets and more practice with tonality and vocabulary. Languages with familiar alphabets and some common vocabulary, such as French, Spanish, and German, are only 24 weeks long.

Each day of classes consisted of six 50-minute sessions with 10-minute breaks and an hour for lunch. Classes are held 5 days a week and each day includes some written homework and the memorization of about twenty to thirty lines of a dialog that will be recited the next day. At the start of class, it would take me more than an hour to do the homework and memorize the dialog. After about three months, I had reduced that to about twenty minutes per day. A typical dialogue to be memorized for the next day might look something like this:

Julie: Les enfants, qu'est-ce que vous voulez emmener pour le pique-nique demain?

Isabelle: Moi, je veux des chips et un sandwich au fromage.

Julie: Et toi, Mike, qu'est-ce que tu veux?

Mike: Moi, je veux un sandwich au jambon.

Julie: Et en dessert?

Isabelle et Mike: Un gâteau au chocolat!

Julie: Bon. Nous allons emmener des boissons fraîches et un thermos de café. Je vais préparer des sandwichs et des chips. Je vais faire un gâteau au chocolat. Nous allons aussi emmener des fruits et des goblets. Vous apportez une glacière et aussi des couverts.

This example was taken from a much more recent DLI manual and is about half the length of the dialogs we would have to memorize each day by the middle of the course.

Each class had about 10 to 12 students. The majority were junior enlisted personnel destined to be linguists in signal intelligence specialties. However, there were also officers and a few civilian DOD personnel. These might be either signal intelligence specialists or officers to be assigned to overseas embassy or military advisor positions. My class had five Navy enlisted personnel, three Air Force and two Marine enlisted and one Army officer. Senior enlisted or officers were also allowed to have their wives take the classes. The senior enlisted person in my class was a Navy first-class petty officer (CTI1) who was taking his second language—Spanish was his first—as part of a reenlistment incentive. I would cross paths with him and other Navy CTs in the class at other duty stations later in my time in the Navy.

DLI classes concentrate on the current spoken language of the target . We didn't spend a lot of time learning the details of the literature and history of France. In this way, the DLI courses are quite different from college language studies. Classes were conducted almost entirely in French. We had textbooks with English translations, but in class we listened to French and responded in French. Early on, the most often used phrase was "Comment dit on…." ("How do you say") followed by an English word or phrase.

Our lead instructor was Madame Low. She was the 40-something wife of an Army colonel stationed at nearby Fort Ord. She was a native Parisienne and always had the attitude, decorum, and dress of an educated upper-class Frenchwoman. She was attractive, charming, and patient with the behavior of her students. I think about half the class had a crush on her!

Madame Low conducted about four of the six daily sessions. The other two sessions were either in the language lab or taught by other instructors with different backgrounds and accents. We had one large, jovial black woman from Haiti who introduced us to the nuances of French as spoken in the Caribbean. Another instructor was a former French Army officer who introduced us to some elements of French military slang which were not acceptable in polite company. Madame Low took our practice with these phrases in stride but referred us back to the original instructor for explanations and further study. We also had an instructor who had emigrated from Algeria and helped us with the accents and vocabulary of French as spoken in Africa. This was handy when I was later stationed in Morocco.

One of the highlights of our last month of school was a class trip to a French restaurant in San Francisco. Madame Low must have done the trip with each class, as she knew the waiter by name and he had been instructed to speak only French. We struggled a bit with the menu but enjoyed the experience. Those of us over 21 could order wine—which we discretely shared with our younger classmates.

Language classes and homework occupied about 40 hours per week at DLI. The rest of the time, we were pretty much free to leave base so long as we showed up for morning formation and classes. Since I had a car, I was in some demand for trips downtown or to the beach on weekends. We often went to Carmel beach for girl watching and a bit of sun. We generally had little luck talking to girls as there was a lot of competition with soldiers from Fort Ord. In addition, 1969 was a time when civilians, particularly in California, were not at all supportive of the war in Vietnam. Thus, there was little chance for socializing with local girls or tourists. Most of the trendy bars in Monterey and Carmel were well above our pay grade. $155 per month didn't go far for food and drink in Monterey at that time. If we really wanted a serious night of drinking, we usually did so at the Enlisted Men's club either at the Presidio or at the Naval Postgraduate School near downtown Monterey. For my underage classmates, the military clubs had the advantage that they would serve 3.2% beer to any serviceman.

One of our evening pastimes was to sit out on the fire escape and watch airplanes landing at the Monterey airport. Our barracks up on the hill was directly below the landing path and the aircraft were only a few hundred feet up as they passed over us. One night we decided to make hot air balloons from laundry bags, balsa sticks and birthday candles. I had learned to make these at UC Davis, where one launch had started a small grass fire when the bag caught fire and fell in a cut-over field. We had to rush out and stomp out the fire and douse it with beer. We didn't have that problem at Monterey, as it was much cooler and more humid. Since the balloons floated off toward the beach, which was several hundred feet lower, none of the balloons were still on fire when they got to the ground. There was one time when a balloon had floated off toward the airport just as a plane came in to land. We wondered if the pilot saw the strange glowing shape—but it was probably lost in the city lights below.

During my six months at DLI, I had one week where I was on mess cook duty in the Army mess hall. The food at that mess hall was not nearly as good as the food in Boot Camp. Our first chore in the morning---at about 5AM, was to lay out bacon on sheet pans to be cooked in the oven. That was the basis of the best lesson I learned in the mess hall: If you need to cook a lot of bacon, it is best done in the oven, not in a frying pan. I spent the rest of the day either washing pots and pans or loading and unloading the automatic potato peeler.

I later ate in much better mess halls at Goodfellow AFB in Texas and the Navy communications station in Hawaii. However, the mess hall at the Monterey Naval Postgraduate school was outstanding. When on weekend desk duty in our barracks, we would find any excuse to make a delivery run to the PG school, so we could eat at their mess hall. The army mess hall had several food poisoning incidents where large numbers of students didn't make it to morning roll call due to gastrointestinal distress. Several years later, when at OSU, I was browsing through a journal of epidemiology at the library. (Why I picked that journal, I don't know.) There was an article about a major food poisoning incident at the Presidio of Monterey. The Army cooks were preparing for a holiday barbecue picnic by pre-cooking some chicken. The idea was to do most of the cooking in the oven the day before, then add barbecue sauce and finish on the charcoal grill the next day. The mistake was in taking the partially-cooked chicken out of the oven and storing it in the walk-in refrigerator in large tubs. The chicken at the outside of the tub cooled quickly. The chicken in the middle of the tubs cooled too slowly and grew lots of bacteria overnight. The day after the barbecue about 40% of the students were incapacitated with food poisoning. The article brought back many less-than-fond memories of the DLI mess hall.

Since the food at the mess hall wasn't very good, we often visited a food truck that showed up in our parking lot at about 8PM. Soon after it arrived in the parking lot, the student who first spotted it would yell out "Roach Coach", then hurry down the stairs to be first in line. In a few minutes, the roach coach had a line of students waiting for their evening snack. This was well before the era of artisanal food trucks that populate our cities today. The menu of the roach coach was basic snacks, the hot dogs, sandwiches and pizza. They didn't do much business in drinks as the vending machines in the dorm had low-cost sodas. The roach coach stayed in business because the snack bar at the EM club was more than a half a mile down the hill. Walking down there and back for a snack could take half an hour out of your study time.

During the last weeks at DLI every class must take a day-long final test. If you didn't get a passing grade, you would have to restart the course if a slot was available. If there was no opening, you could be reassigned to a different rating and could end up in the fleet in a crummy job. Stress levels were high, but everyone in my class passed with acceptable grades. All I remember about the test is that one part was to listen to and summarize a French Language news broadcast. The one we got was about a plane that crashed into the ocean and where many people survived the impact, but later drowned.

Once I had mastered the daily dialog and homework, I started finding other things to occupy my evening hours. I did watch a bit of TV in the barracks day room, but that soon bored me. The only TV I remember from DLI is watching the live broadcast of the first landing on the moon on July 20th.

Since I was bored with TV, I started studying for advancement in rate in the evenings and doing some electronics experiments on my desk. One experiment was an audio oscillator. I found out that high-frequency sounds can be very annoying and hard to localize. I would sometimes open the door (we all did that to get a breeze through the room on warm evenings), and start the oscillator. Soon there would be people in the hall asking, "What's that sound?" I was pretty good at shutting it down before anyone found me out. I did have to limit these pranks to times when my roommate was out of the room, as he would haven ratted me out in an instant.

It was an oddity of the Naval training system that, while the duties of a Communications Technician were mostly classified as Secret and required Crypto access, you could study for the CTI3 rank using only a training manual classified Confidential. After some discussion with the CTI1 Gillis, the first-class petty officer in my class, I signed up for the advancement correspondence course and got the manual. I had confidential access as a normal part of the training after a quick check of criminal records (The NAC or National Agency Check) in boot camp. At that time, my full background check was still in progress.

The CTI 3&2 Manual was mostly about the organization of the Naval Security Group and basic electronics and radio propagation. There was really nothing about the real day-to-day work of a signal intelligence collection operator or linguist. We got all that training at our next school, the Radiotelephone Operator School at Goodfellow Air Force Base in Texas.

I finished the correspondence course and qualified to take the Navy-Wide advancement test in October of 1969. Another requirement was that you had to have six months rated as Seaman (E3) by the time of the exam. I met that requirement since I had been rated at E3 from the end of Boot Camp. I passed the test with a score high enough to be selected for advancement to Petty Officer Third Class (E4 or CTI3) at the time I graduated from DLI. By that time, my background check had been completed and I was granted a Top Secret Crypto clearance. Getting that clearance was a pre-requisite for the next training school.

As he "pinned on my crow" with a punch to my newly-attached Petty Officer Third Class insignia, CTI1 Gillis told me "Borgerson, you got it dicked now—no mess cooking at Goodfellow or your next duty station." He was right, and the lack of mess duty and the extra $60 per month drew a bit of envy from my classmates.

I graduated with honors from the French Language course at DLI in October of 1969, ranked third in a class of 36 (there were two other sections of about 12 students). After a week of leave, I headed for the Radiotelephone Operator course at Goodfellow Air Force Base in San Angelo, Texas. I later discovered that after some time in the 1980s graduates from the DLI basic course could be awarded an Associate degree in the language studied. It seems that six months at DLI is the academic equivalent of two years of undergraduate college work. Five years later, after working as a military linguist for four years, I sat in on a few junior-level conversational French classes at Oregon State. I found that college French majors may know the literature and the history, but they sort of suck at everyday conversation, and their knowledge of modern slang and profanity was minimal. Six months of immersion classes and a few thousand hours listening to and translating radio broadcasts gives you quite a different skill set.

Figure 3 View from DLI dorm fire escape.

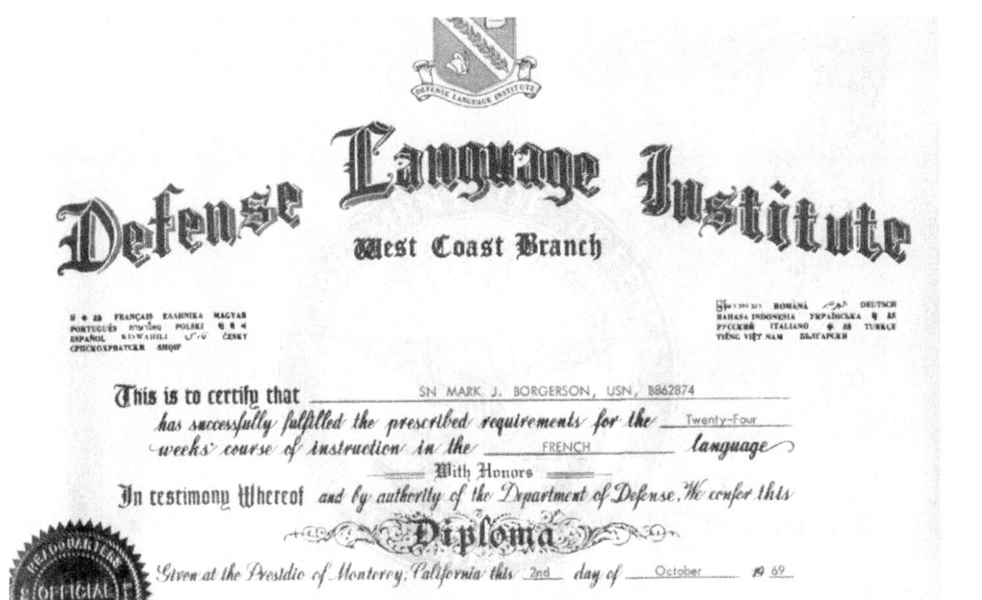

Defense Language Institute

West Coast Branch

This is to certify that SN MARK J. BORGERSON, USN, B862874

has successfully fulfilled the prescribed requirements for the Twenty-Four weeks course of instruction in the FRENCH language

With Honors

In testimony Whereof and by authority of the Department of Defense, We confer this

Diploma

Given at the Presidio of Monterey, California this 2nd day of October 19 69

LIEUTENANT COLONEL, USA
ADJUTANT

COLONEL, USA
COMMANDANT

OFFICIAL

Figure 4 DLI Graduation Certificate

Figure 5 CT Rating Badge.
This one is for CTI2, which I earned just after I arrived at OCS

Chapter 4 Radiotelephone Operator Training

San Angelo Texas is a city 120 miles west of Austin. In 1969 the population was about 63,000 and it was home to the Angelo State University and Goodfellow Air Force Base. Goodfellow had been a pilot training base during WWII, but was primarily a technical training base in the late 1960's. I have no memory of how I got there in the Fall of 1969, but I presume it was some combination of planes and buses from my leave at Arcata. The area seemed a bit cold and bleak after Monterey. I do remember that the food in the Air Force mess hall was a great improvement over that in Monterey. The fact that I was exempt from mess hall duty as a newly-promoted petty officer was also pleasant.

The training schedule had us learning the basics of radiotelephone intercept operator duties. One of the requirements was that you had to get a minimum proficiency in typing. I was fortunate that typing lessons from my mom had gotten me well past the requirements. We had to learn how to operate tape recorders and radio receivers, as well as the organization of the military cryptologic services that were the primary collection resource for the National Security Agency.

Here, we were introduced to the R-390 High Frequency radio receiver. This receiver used vacuum tubes and mechanical tuning to cover the radio frequencies from about 500KHz to 30MHz. It was exceptionally stable and versatile and remained in use in the military from the 1950's to 1990s.

Since all the students now had full Top-Secret Crypto clearances, we finally learned exactly what we would be doing for the rest of our military careers. We were introduced to the fundamentals of signal intelligence, and communications intelligence (COMINT) in particular. Our job was to be the interception, recording, and translation of foreign radio communications. We would be working at one of several dozen US and overseas military bases which hosted the service cryptologic agencies (Naval Security Group, Army Security Agency, Air Force Security Service, and Marine Corp Cryptologic Support Battalions).

Our class was a group of the most recent graduates of DLI. It included members from all the services. Most of my class from DLI was in the same course—except the officers and senior enlisted who had already had equivalent training.

Two of the main subjects of training were TEXTA and TECHINS. TEXTA stands for "Technical EXtracts of Traffic Analysis". Or, from an NSA Document:

> "a. TEXTA - A term denoting case tetragraph and number information which includes all pertinent details of target communications circuits. The term itself is an abbreviation for the phrase "technical extracts of traffic analysis".

In the 1960s and 1970s, TEXTA took the form of a large set of index cards, sorted by target nation and target type. Each card had a 4-character country and target type and a 4-digit circuit number. One case number was something like this:

(I think—some of the details are a bit hazy.) 6101 may have referred to a link between the Pacific test center and an outlying weather reporting station.

The card also included the stations communicating on this circuit and the times and frequencies that they used for communications. When we were assigned intercept duties we were assigned a prioritized list of TEXTA entries. We would monitor the frequencies and intercept communications that we received. Some circuits had high priority and would always have a receiver tuned to that circuit. An intercept position usually had at least two receivers, one for the priority circuit and another to cycle through the lower-priority circuits. Sometimes we had a recorder for each circuit and other times we had to switch recorders to the highest priority active signal. In addition to the voice signal from the receiver, we recorded operator comments such as the time of intercept, the TEXTA designator, and the operator name and rank. We also kept a paper position log noting the time and general contents of the intercept.

We had about 7 hours of class per day with weekends off. There was an enlisted club that set up weekend dances and mixers with young ladies from the local college. The people of San Angelo were very friendly to the military and the events were well attended. The friendliness was probably due to the character of Texans and the fact that the locals probably realized that the Goodfellow students were the "good boys" of the military. (Bad Boys didn't get security clearances.)

One of the premier social events of the fall was the Navy Day Ball. We all worked hard to find dates for the ball. I managed to get a date, as did one of my DLI classmates, Steven H.

As I recall, my date was named Dana and was a junior at Angelo State. We went out on several other weekend dates as well as the ball. The ball was the height of the social life for the class. I'm not sure where it ranked for the young women, but they seemed to enjoy the dancing, food, and drinks. I think all involved realized that it would be very unusual for any long-term connections to occur. The women knew as well as we did, that the soldiers, sailors, and airmen would only be around for a few months, then depart for a new life overseas.

Early in December of 1969, I finished my course at Goodfellow and departed for the Naval Communications Station in Morocco.

Figure 6   R-390 HF Receiver front panel

THE U.S. AIR FORCE SECURITY SERVICE SCHOOL
6940th SECURITY WING
GOODFELLOW AIR FORCE BASE, TEXAS

Certifies that

CT3 MARK J BORGERSON, B862874

has successfully completed the

RADIOTELEPHONE TRAINING COURSE ALK 20331-Z

and is herewith awarded this

CERTIFICATE of TRAINING

ROBERT P CRAIG, Col, USAF
Commandant

4 December 1969

Figure 7 Radiotelephone Operator Certificate

Chapter 5 Traffic Analysis and Metadata

Until very recently, people outside the SIGINT community were really not aware of the amount of information you can gain about a target without actually reading their messages. Traffic Analysis and the TEXTA it generated were a major source of intelligence for the NSA and the Service Cryptologic Agencies during the cold war. Even when the SIGINT community could not decrypt the body of a message, it could, by keeping voluminous records of who was sending messages to who, learn a lot about the structure of a target organization. Careful analysis of the flow of message traffic between stations can allow you to determine who is in charge and who is reporting to whom.

When there are regular patterns of communication, the volume of traffic can allow you to establish a baseline traffic volume. When there have been past sudden increases in message traffic you can correlate those increases with other intelligence observations and learn to predict activity patterns that are preceded by traffic volume spikes. If you can map communications identifiers to particular target units, you may be able to detect changes in the position of mobile units by using radio direction finding.

In the last few decades, traffic analysis has become both more difficult and more complex. Many target communications are now transmitted over optical fiber and are no longer easily intercepted. Widespread use of computer networks provides both more secure communications channels and more vulnerability to malware. Classic traffic analysis as practiced in the 1970's is probably no longer a major source of information about either the Russia or China, but it may still be valuable in analyzing terrorist or third-world communications. Much of modern traffic analysis is probably done by computers, and the NSA still hires many mathematicians and statisticians who can attack modern traffic analysis problems.

Recent revelations about the NSA have shown the increasing degree to which both intelligence agencies and law enforcement collect and exploit the metadata (or data that describes the type of data in a message). The collection and exploitation of metadata such as the time, location, and duration, and recipient of a cellular telephone call, is much less strictly controlled by the Constitution and privacy laws than is the actual content of the conversation.

The NSA has recently brought on line a very large data storage facility in Utah. There is speculation that the facility could store all the metadata for every telephone call and internet packet leaving or entering the USA in a year or more. Whether it stores the actual internet data or digitized telephone conversations is open to debate.

Some internet security experts say that you should assume that the NSA is recording everything you send or receive over the internet—especially if you communicate outside the borders of the USA. If you don't want them to read your emails, you need to use strong public key encryption. Even that may not be enough very soon, as the NSA is likely very interested in developing quantum computers that may multiply by orders of magnitude their ability to defeat current encryption.

Now, as was the case back in the 1970's, the biggest challenge for the NSA will be how to effectively analyze the massive amounts of data and metadata it can collect and store. Advancements in machine learning and machine translation may help, but the intelligence community will rely on trained analysts to detect patterns and exploit them for many years to come.

Chapter 6 Sidi Yahia Morocco

In 1969, Morocco had been independent of France for thirteen years. The king and country maintained good relations with the US and NATO Countries and has been a consistently moderate North African Muslim country. The US had a large Air Force Base at Sidi Slimane, northeast of Kenitra until 1963. The closing of that base was prompted by political tensions and the Air Force's ability to accomplish their mission from less sensitive bases.

Our nominal home base in Morocco was at Port Lyautey, or Kenitra. It was home to a US Naval air station since WWII days. It was also a training base for instruction of Moroccan Air Force pilots. It provided a diplomatic cover for the US Naval communications stations at Sidi Yahia and Bouknadel. Naval personnel were not allowed to travel outside the US bases in uniform to minimize our footprint in the local area. We were bussed from our station at Sidi Yahia to Kenitra for liberty, as there were no recreational opportunities in the small village of Sidi Yahia near our station. The on-base recreational facilities consisted of an enlisted club, swimming pool, baseball fields, and theater. The Kenitra Naval station had a larger club and there were bars and shops in the city. The bars and shops had grown to accommodate the influx of airmen from Sidi Slimane and were desperate for new customers after that base closed. The local area was starting to become a destination for European tourists, especially in the winter. The Kenitra area produced a lot of oranges and eucalyptus timber. The eucalyptus wood was converted to charcoal cooking fuel in large earth-covered ovens. We would often pass a dozen of these smoldering mounds on our bus trip to Kenitra.

When I arrived, I was assigned to one of four watch sections that [redacted] [redacted] around the clock. We had what was called a "2-2-2-80" watch rotation. It went like this:

Eve Watch	4PM to Midnight
16 hours off	
Eve Watch	4PM to Midnight
8 hours off	
Day Watch	8AM to 4PM
16 hours off	
Day Watch	8AM to 4PM
8 hours off	
Mid Watch	Midnight to 8AM
16 hours off	
Mid Watch	Midnight to 8AM
80 hours off	

This watch schedule had some good points and some bad points:

GOOD	You only worked 48 hours of each 192-hour rotation, or an average of 6 hours per day.
GOOD	You got the equivalent of a long weekend once each 8 days.
GOOD	The Navy offered a midnight meal, "Mid-Rats" from 11PM to 1AM each day.
BAD	Day shift meant eating a bag lunch, as there wasn't time to get to the mess hall and back.
BAD	The 8 hours off between Eves and Days and Days and Mids left little time for anything but a quick meal and sleep.
BAD	After Mid watch, you had to sleep during the day and the barracks was hotter and noisier.

My working area at Sidi Yahia was in the receivers building, which is about 1.6 miles southeast of the main base. This building was concrete, two stories tall and had no windows. It had excellent air conditioning to keep the hundreds of radio receivers and pieces of crypto gear cool. We were bussed to and from the building for each watch. It was a bit disconcerting to see a group of sailors waiting for the bus on a 100-degree afternoon. They would all be in dungaree pants and long-sleeved shirts, carrying dungaree jackets. I had been advised to bring my jacket for my first watch but wondered why. I found out when I walked into the receivers building. There was an outside vestibule that was about 80 degrees. Along with my watch section, I put on my coat, and walked into the receiver room. It was about 70 degrees inside—a temperature I thought very comfortable. When I sat down at my operating position, I found out that the air conditioning vents were under the equipment racks and cold air was blowing out of the cracks between pieces of equipment. That air temperature was about 55 degrees!

A normal day or eve watch was spent at a radiotelephone position monitoring naval traffic.

At the end of each day the magnetic tapes were collected

During mid watches we would do office chores like updating TEXTA, translations, and housekeeping. I remember that floor mopping and waxing in the office area was always more time consuming after swivel chair races around the area left skid marks on the floor.

After I had been at Sidi for about 6 months, I got upgraded from junior operator to translator

It was more challenging, but more work. It left little time for exploring the airwaves for signals lower down the priority lists. It also meant less time to covertly tune a receiver to Radio France for music and "language practice".

The receivers building had a Marine guard at the entrance who would check your badge before allowing you to enter. Other Marines with guard dogs patrolled the antenna field at night. The high tensile strength copper wire used in the antennas was prized by local artisans who would convert it to hammered copper jewelry. Occasionally, an enterprising thief would climb an antenna support to cut loose some wire. If he was detected by the guards, they would sit below him and smoke a few cigarettes while the dog snarled at the thief. After a few cigarettes, they would call in the incident and a jeep would arrive with the sergeant. They would discuss the situation while waving around their firearms, then back off and yell "La!" (Get out of here!) and let the thief escape. There was too much paperwork in the interaction with the local police if they apprehended the thief. The thieves often came back for another try because they knew the Marines would let them go. If the local police caught a thief on a telephone pole, shots WOULD be fired.

The enlisted barracks were one-story poured concrete buildings with minimal ventilation and no air conditioning. We had two and four-man cubicles with head-high partitions or curtains and two-high bunk beds. Since there would usually be men from one or another watch section sleeping, the interior was generally dimly lit. There were bathrooms and laundry facilities at the end of the building. We would generally get a watch section of 10 to 12 people together and hire a local houseboy. For about $15 to $20 per week, he would sweep and mop our section of the barracks, wash our clothes and do our share of the bathroom cleaning.

The barracks would get very warm—over 90 degrees—on summer days. That could make sleeping coming off a mid-watch very difficult. The internal partitions limited the air circulation and just about everyone had a small fan blowing at their bunk. When the fans were all going, that added about 3000 Watts of heat energy to warm the air even further. The partitions between 4-bunk cubes were only about 6 feet high, so they didn't block much sound. There were quiet hours from about 8PM until about noon, but after that you could hear stereo equipment from other cubes. The conditions in the barracks were one of the reasons that I applied to Officer Candidate's School—it would get me out of Morocco a year ahead of my normal rotation schedule.

The base Marine Corps security detachment held security exercises on base about once a month. These exercises usually involved a Marine playing 'intruder' being pursued by the Marine detachment. For reasons involving Navy-Marine rivalry, the intruder would usually be cornered right outside the CT barracks and a fire fight with blank ammunition would occur. These exercises usually happened at about 10AM---just about the time you got soundly asleep after a mid-watch.

The mess hall at Sidi Yahia wasn't very good. It wasn't as bad as the Army mess hall at DLI, but it suffered from the fact that it was at the end of a long and intermittent supply chain. As a petty officer, I didn't have to do mess duty, but my friends who did told me about some of the problems the cooks at had to face.

All the American staples arrived by ship to Kenitra and the ships only arrived about once a month. That meant we didn't get fresh corn or other normal American vegetables. We did have some fresh local vegetables, but due to sanitation problems, they all had to be disinfected with a dip in diluted bleach solution. The cooks kept the mess hall clean and we didn't have food-borne illnesses. However, the menus seemed more like those you would find on a ship that had been at sea for several weeks. (A few years later, I was on a ship that spent more than six weeks between port calls, so I speak from some experience.) However, like a ship at sea, you could always count on a good breakfast. We got fresh eggs, lots of bacon and sausage, and pastries baked daily.

In addition to the mess supplies, the ships brought in the goods for the base exchange. It was easy to judge the time since the last ship by checking the shelves at the exchange. After a few weeks the more popular snack items and toiletries would start to disappear from the shelves. These vagaries of the supply system led to a bit of a hoarding mentality and the lockers in the barracks held a lot of snacks and sodas.

The supply ships also brought in the beer and liquor for the clubs on base. They brought in a LOT of beer, and I don't remember the EM club ever running short. The beer was also very inexpensive in the clubs. You could get an 8-ounce paper cup of beer for a nickel at happy hour. On Fridays, you could get a quart of Schlitz or Pabst Blue Ribbon for a quarter. Those of us from the western US always complained about the lack of Coors—but the company then, as now, required that the beer be shipped cold. That wasn't going to happen on the ships to Kenitra. We would often get cans or bottles of beer that had gotten too warm for too long and had gotten 'skunky'.

The EM club also had a seemingly endless supply of pizza and hot dogs. These could be prepared with canned ingredients, so they didn't taste of bleached vegetables. When coming off day watch, we would often skip dinner at the mess hall and head to the club for pizza and beer. You could usually get the same food at the mess hall—but the experience lacked the ambience and beer of the club.

About once a week, the EM club would hold a dance, for which they would import young women from various girl's schools in Rabat. The women were generally of college age and were a mix of Moroccans and the daughters of expatriate Europeans. The goal for the women was to gain some exposure to American culture and some practice with conversational English. The goal for the sailors was the same as for any sailor with limited exposure to English-speaking young women. Whatever the sailor's goals, the interaction was generally limited to food, drink, dancing and conversation as the women were not allowed to leave the club before the bus back to Rabat.

When there was room on the bus, the driver would sometimes allow one or two sailors to ride back to Rabat with the women. I and a couple of other CTIs convinced the driver that the 1-1/2-hour trip to Rabat would give us a good chance to practice our conversational French. The girls who managed to stay awake seemed to enjoy the trip as they got to laugh at our French vocabulary and accents. The guys enjoyed the trip as they got a chance for some language practice and an opportunity for a bit of kissing and groping. The trip back to Sidi Yahia at about 2AM was much less lively, but we usually had at least one or two people awake to talk to the driver and help him stay alert. There usually wasn't much traffic outside the cities at night, but he did have to watch out for stray goats and sheep. For the driver the main incentive for allowing the sailors on the bus was the company on the return trip.

Kenitra, which was our primary liberty destination, was a fairly new city, having been established in 1913 by the French colonial government as a port to rival Casablanca. In 1970, the population was about 100,000. There was a well-developed downtown area with shops and numerous bars and night clubs. We were dropped off in this area by the liberty bus from Sidi Yahia. Buses ran about every two hours from 8AM until 1AM. If you missed the last bus, there were some all-night bars and ladies of the evening who would provide overnight accommodations.

Normal liberty in Kenitra started with a 25-minute bus ride from the base into town. When we got off the liberty bus, our first stop was usually at our favorite street food vendor. The most popular dish for the Americans was brochettes. These were skewered chunks of spiced lamb or beef grilled over charcoal and served on flatbread. As I recall, you could get a pretty good meal for about 3 Dirhams. At that time, the exchange rate was about 5 Dirhams per US dollar. If you wanted something fancier than brochettes and a drink with your meal, you could have local Moroccan dishes or a limited selection of French and Spanish dishes in a café or bistro. A good meal would cost 10 to 15 Dirhams. A Heineken beer would be about 5 Dirhams. There were other cheaper Spanish beers, but we usually drank Heineken as it was uniformly of good quality. Coca Cola and Fanta sodas were also available. We had been warned to stick with those brands and not ask for ice to avoid food-borne illnesses.

The Moroccan Gendarmes were pretty good at keeping order and would strongly discourage any hassling of the Americans. The Navy dollars spent in town were a valuable resource and local merchants didn't want anything interrupting commerce. Kenitra also had a train station with regular north and south departures. You could take a one-hour train ride south to the capital of Rabat for the equivalent of a few dollars. Casablanca was about 50 miles further south. I made several trips to Rabat for sightseeing and one trip to Casablanca with Dan Hogan, with whom I had attended DLI. Almost all the merchants and bar workers spoke French, so communications were no problem. It was awkward to be speaking the language, since we seldom did that at work. There, we listened, took notes, and translated to English.

Many of my friends explored further afield to Roman ruins and Moroccan historic sites in cities like Fez, Meknes, and Marrakech. I didn't do any of that, since I soon got involved in the little theater group that staged plays in the base theater. I also qualified as a projectionist and earned some extra money showing movies in the theater. The most heavily attended movie of the year was Barbarella---which featured a few nude scenes with Jane Fonda. After the first few showings, I watched the opening credits, then read a book for the rest of the movie. Movies were shown at midnight on a few days of the week so that watchstanders who were working eve watches could attend. These late showings were more popular in the summer as the theater had poor air conditioning and would be cooler later in the evening.

While I was at Sidi Yahia, I worked on two plays: "The Pajama Game" and "Bus Stop". I did a lot of the set design and construction for the latter play. Sidi was a small base and there weren't a lot of people interested in working in the theater during the heat of the day. While I was working in the theater, I met Mark B, a CTM3 maintenance tech. He and I became best friends and later met for a few days of beer drinking at the Octoberfest in Munich.

I received orders to report to OCS on the 9th of October. Just before leaving Morocco for OCS, I took the Navy-wide advancement exam for CTI2. I wouldn't find out the results until my next duty station as I was granted 25 days leave plus travel time on my way to Newport, Rhode Island for OCS. I had my mom purchase a 21-day EuRail pass and send it to me for my travels in Europe. I left Kenitra on a train to Tangiers on the 24th of September 1970. I had shipped most of my uniforms and personal goods ahead to Newport. I had my civvies and one set of blues in my trusty Navy sea bag. (That sea bag is still around—having been used on many charter boat and research vessel trips.)

Figure 8 Aerial view of main station at Sidi Yahia

The enlisted barracks are the long one-story buildings in the center. Officer and CPO housing is at the upper right.

Chapter 7 European Leave

After taking the ferry from Tangier to Tarifa, Spain, I caught an overnight train to Madrid. I spent just one day there before taking another train to Amsterdam. I postponed activating my EurRail pass until I left Amsterdam.

When I arrived in Amsterdam, it was late in the day. I visited the nearby tourist office and collected maps. It happens that the famous red-light district is just south of the main station, so I took a quick early-evening tour on the way to my hotel. Over the course of the next two days, I visited the Rijksmuseum to see the famous "Night Watch" by Rembrandt and walked around the canals taking pictures.

I left Amsterdam on a train to Cologne, which has a famous cathedral. I visited the cathedral, but my main goal was to visit the PhotoKina World Fair of Photography which was in progress. I toured the vendor floor and took some pictures of various models posing for the attendees. After a day in Cologne, I boarded a train for Munich.

The trip to Munich started out with a very scenic route along the Rhine river. After a few train changes, I ended up in the forests of southern Germany. My high school and college German was sufficient to allow me to get around the cities without problems. When I got to Munich, I found a hotel and left a note at the local American Express office for my friend Mark B. Mark was taking a week of leave and we planned to meet for the Oktoberfest. Mark joined me at my hotel and we made plans to visit the Oktoberfest the next day. I wandered around the city that morning. In the afternoon, Mark and I headed to the Oktoberfest grounds.

My main impression of the Oktoberfest was "LOTS of people and lots of beer. Each major brewer had a building with tables music, food and beer. The Lowenbrau building had a large animated lion raising a stein over the entrance. Inside, we watched waitresses carrying six one-liter steins of beer at a time; three in each hand. There were kids barely large enough to see over the edge of the table sipping beer poured into smaller steins by their parents. At one point, a man of about 70 years got up onto a table in a Kaiser Wilhelm helmet and army uniform and started marching back and forth to cheers from the crowd. Apparently, Kaiser-era military regalia was acceptable—but Nazi-era uniforms were not. Mark and I drank a lot of beer—Mark B a bit more than me. By the time we left at about 11PM we were well under the influence. That made us about average for the crowd at that hour. I remember that Mark was unsure of the direction to the hotel, and I was a bit unsteady on my feet, so I walked behind him with my hands on his shoulders, yelling out "Links", "Geradaus", or "Rechts" (Left, straight ahead, right). We made it to the hotel without any incidents that left bruises or memories. Next morning, we slept in, then headed out to the Rathaus for beer and brats for lunch. Alas, those recuperative abilities are but a fond memory. Later that day, Mark got on a train to head back to Morocco. I walked around the city in the afternoon and visited the Munich museum of Science and Industry. That evening I caught a night train for Switzerland.

I took the train from Munich to Interlaken, Switzerland. Interlaken is at the foot of the Swiss alps and offers easy access to such tourist sites as the Matterhorn and Jungfraujoch. I got a hotel with a small balcony overlooking a park. Over the next two days, I walked around the town and took the cog railway to Jungfraujoch—a scenic point high in the alps with great views over the mountains and valleys.

The cities and train rides in Switzerland lived up to my expectations. The cities were immaculate, and the trains seemed to be routed through well-tended parks. No litter, junked appliances, or trashed cars were visible. The only US area with such neat appearance that I've seen from a train was when I was going through Iowa on a trip from Denver to Chicago. Perhaps Iowa has a law that only Swiss immigrants may purchase property within 200 yards of an Amtrak line!

While waiting for my train from Interlaken to Rome, I met a young American woman who was a recent graduate from an east-coast college. Her name was Joanne, and she was doing a tour of Switzerland and Italy before going to Paris to meet her uncle and aunt. She was a bit concerned about how men in Italy would behave with a single American woman. With an abundance of chivalry and a year-long dearth of contact with single American women, I proposed that we might travel together while in Italy.

She thought it over until just north of Rome and decided my company would be less irritating than traveling alone.

We found a youth hostel in Rome and spent about three days there. There were about a half-dozen other English-speaking young people staying there. They gave us hints about what to see in Rome and where to find good food at low cost. We spent three days touring Rome. We saw St. Peter's Basilica, the Vatican Museum, the Trevi Fountains and Colosseum. In the middle of our tours, we took a bus to the beach at Ostia with three or four of the other people from the hostel.

As we left Rome, we parted ways for a bit. Joanne was off to Paris with a stop in southern France. I headed to Florence for two days before taking an overnight train to Paris. On this leg of the trip I paid a supplement to my EurRailPass to get a couchette: a small bunk in a 3-high tier with a privacy curtain. It was well worth the supplement as I arrived in Paris well rested. Joanne had given me the name of the hotel in Paris, where she would be staying with her uncle. It sounded way above my budget limit, but she would be there for at least a few days before she headed to Bordeaux with her uncle and aunt for a tour of the wine-growing regions.

When I arrived in Paris just after noon, I found a hotel and spent the afternoon sitting on my small balcony drinking hard cider from Normandy and chewing on a baguette. The next day, I wandered around the city, including a trip to the top of the Eiffel Tower. Thanks to six months with Madame Low and a year in French Colonial North Africa, I felt comfortable moving around the city.

That evening, Joanne and I had dinner with her aunt and uncle in a very nice restaurant. Joanne must have given me a good grade, as we had some expensive wines and a great meal on her uncle's tab. The meal cost more than I had spent on food over the last 4 days. I'm sure her uncle was more impressed with my language skills than my knowledge of French wines—since I knew nothing about French wines. Thanks to a few tours to Napa Valley during my senior year at UC Davis I had some rudimentary knowledge of California wines. He was appreciative of my attendance at the premier school of oenology in the US, but probably wished I'd studied wines instead of analytical chemistry. Nonetheless, he was interesting and cultured and Joanne and I had a good time. At the end of the evening, Joanne and I exchanged addresses and promised to get together should we be in the same city at the same time. That actually happened about five months later, after OCS.

I spent part of the next day walking around Paris, then made my way to the Charles de Gaulle Airport to catch my Navy-booked flight to New York. I arrived at the airport with just about $35—just enough for bus fare from New York to Newport, and OCS. I then found that I had to pay about $25 in airport taxes that were not included in my travel voucher! I paid the tax and spent the flight to New York wondering how I would get to Newport without finding a Navy base and pleading poverty. That wouldn't look too good on the record of a responsible young man aspiring to become a Naval Officer! I had bus fare to Manhattan and spent the day wandering around the city and thinking about my predicament. Worst case, I could get to the Brooklyn Navy Yard and plead poverty. My solution was to find a pawn shop and hock a $120 Canon telephoto lens I had purchased duty-free at the Navy Exchange in Morocco. The US value of the lens was over $200, and I got about $55 at the pawn shop. That was enough to cover dinner and one night in a cheap hotel, as well as bus fare to Newport. I arrived late in the afternoon on October 6th. One good thing about Navy bases: they're open 24 hours a day and if you've got orders, they'll feed you and give you a place to sleep.

Figure 9 Gelato in Rome

Chapter 8 NAVOCS Newport R.I.

I returned to the Quarterdeck (front desk) of the Navy Base the next morning to finish my check in for OCS. I found that my class wouldn't start for a few days. I also found that I had passed the test and been promoted to CTI2 (E-5). For the time being, that had no real effect, as all officer candidates are paid at the E-5 rate. I found the disbursing office and collected a few dollars of pay---just a few, as I had taken a month of advance pay just before leaving Morocco. Still, I had a berth in temporary quarters, knew where to find the mess hall, and my uniforms had arrived. I figured I was set for the next four days. Since I was an honest-to-God Petty Officer with 22 months of Naval service, I was assigned to a lead group of other early arrivals in various chores—among them assembling about 100 new lockers to go into a new barracks building. The CPO in charge of our detail saw no profit in pissing off a group of future officers, so he just told us what to do and that we were to work from 8AM to about 4PM with time off for lunch. With my vast experience in matters Naval, I figured he was saving his own division for work that required actual Navy experience---like brewing coffee and fetching the proper selection of donuts to get him through his morning paperwork. We worked at what we deemed to be an acceptable unsupervised Navy activity level and managed to finish the required number of lockers in a few days. While working, we shot the shit and speculated on what OCS would be like. One of the other people in the group was a Sonar Technician with about the same service time as me. He was also designated to transfer to the Naval Security Group after commissioning, and we grew to be good friends. His name was Steve Pennick and we would share a beach house at our next school in Pensacola, Florida.

When the time came to form up our OCS class, I was assigned to Company Charlie in class 7104. The '71' indicated our graduation year and the '04' specified that we would be the fourth class of that year. Generally, one class graduated each month. Each class had 16 weeks of training, but our class took about 18 calendar weeks as we were granted two weeks of holiday leave at Christmas time.

Each company had a junior naval officer as its company commander. After a few weeks, the company commander would appoint Officer Candidate commanders to be company leaders during drills, marching to classes, etc. My company commander was one of just a few black officers at OCS. He had just returned from a year stationed ashore at DaNang in Vietnam, where he was a supply officer. A year of hazardous duty earned preference points for future assignments and he had picked NAVOCS to be near family on the east Coast. According to him, DaNang really was hazardous duty—he lived through several mortar attacks on the base. One of them destroyed his hut and several hundred dollars' worth of new stereo gear.

The first thing I discovered about OCS is that it was much more demanding, both physically and mentally, than boot camp. Our physical training in boot camp had been curtailed by a lot of rainy weather that kept us from outdoor activities and RTC San Diego didn't have a lot of indoor athletic space. No such luck at OCS. We started our training the first week with a lot of running. Even after a lot of walking in Europe, a year at a desk job had left me ill-prepared for the level of performance expected. Classmates who had been athletes in college had no problems—it was probably easy compared to college sports training. Luckily, there were enough non-athletes in the company that I wasn't alone in needing some remedial PT. NAVOCS also had plenty of indoor gym space so training wasn't interrupted by bad weather.

Of course, one of the first things that the Navy taught OCs (Officer Candidates) was how to fold your clothes, clean your bathrooms and mop and wax the floors in the dorm. I'd already had months of training and experience in those areas and was immediately appointed as chief electric buffer operator! We didn't have to wash our own clothes. Instead we got a laundry allowance. Laundry trucks would line up on the road outside the residence hall several times a week. You took out your clothes, and they came back fresh and clean a few days later. This was necessary, as the wool winter uniforms required dry cleaning and pressing—skills completely foreign to most 24-year-old men. Interestingly, OCS was about the last time I wore wool dress blue uniforms. While I was in Hawaii, the uniform of the day was khaki slacks and a short-sleeved khaki shirt. Soon after arriving in Hawaii, I invested in permanent-press slacks and shirts and reverted to doing my own laundry. While this was less costly, there were occasional stressful Sunday evenings when I found out I was one item short of a clean uniform for Monday morning. I still have some nightmares about finding a uniform shop to replace a torn or stained shirt.

Some of the early academic courses were easy for me—I had absorbed a lot of Naval history and organization through Sea Scouts and my Naval service. Other courses, such as inshore piloting and celestial navigation were more difficult. They required a lot of arithmetic proficiency and attention to detail. It's important to remember that navigation at that time did not benefit from the GPS, calculators, computers, and plotters that are ubiquitous today. I managed to work my way through those courses—never at the top or bottom academically. A few years later those courses would turn out to be very valuable as I sailed around San Francisco Bay. A few years after that, I found that charter boat agencies would accept Naval Officer training and a short exam as proof of competence to charter a sailboat or trawler yacht.

The piloting and navigation problems we had to solve often took place in southern Alaska. That area has several characteristics that make it a good area for piloting exercises: narrow, winding passages, large tidal changes and strong currents, lots of islands and navigational markers. A typical problem might be something like this:

You are aboard a destroyer escort in Clarence Strait proceeding on course 110 true at 12 knots. The time is 1900 local on December 12, 1970. The quartermasters have reported the following bearings:

A white flashing light with a period of 4 seconds bears 045T
A white flashing light with a period of 2.5 seconds bears 281T
The radar bearing to the Western edge of Cape Fox is 130T

Your goal is to arrive at Ketchikan Harbor on 13 December at 1000Local.

Specify your true courses and speeds to arrive at the destination as planned. Consider tidal current set and drift and assume a minimum enroute depth of 100 feet. Approach no obstruction closer than 500 yards. Maximum allowable speed is 12 knots.

Answer the following questions:
What will be the set and drift as you pass Twin Islands Light?
At what time will the Angle Point light bear 045T?
At what time will you be abreast of channel marker 3 at the harbor entrance?
At what time will marker "CR" bear 045T?

A problem like this would take us about a week to solve. The problem required you to look up the characteristics of various lights, calculate the tidal currents using current tables, and plot your course on your copy of the appropriate chart. You had to account for the fact that a 300-foot destroyer doesn't turn on a dime—at 10 knots the ship might advance along the previous course for a hundred yards before steadying on the new course. The navigator on a real destroyer would be expected to have a solution in about 4 hours. Of course, he would have the help of a chief quartermaster and notes from previous passages. We could present trial solutions to our instructor. Sometimes they would come back with notes like "On the rocks at 0200! Set and drift incorrect."

In 2010, I made a passage from Ketchikan to Bellingham Washington in a 37-foot trawler with a maximum speed of 11 knots. I was with a friend and business colleague who had gone to OCS at about the same time as me. We had a bit of fog in Dixon entrance and solved much the same kind of navigation problem. Our situation was much simpler as we had GPS, radar and an electronic chart plotter. We didn't have to translate bearing from the bridge crew into our boat's position. However, we did check radar bearings to land features to verify our position. We also had to contend with a lot of fishing boats and drift nets along the way . We both remembered the stress we felt as we practiced navigating a much larger ship through the same passages in the dark.

For more real-time practical piloting and maneuvering training we spent several days in a bridge simulator based on high tech computers and displays for the era--about one-quarter the computing power of the first IPhone. We also spent several days on the water in teams of about eight piloting 80-foot yard patrol boats (called Yippies for their YP designation) around Narragansett Bay.

Unlike most of my fellow OCs, I wasn't too motivated by class standing or academic performance. For most of the OCs, standing and performance were the key to getting their choice of follow-on schools and duty stations. Since I was one of just a few OCs already designated to become a Cryptologic Officer, class standing was not a big motivator. I knew I would go to the Cryptologic Officer Indoctrination course in Pensacola, then on to a shipboard or shore NSG department. My language skills would be a major factor in my assignment—and those wouldn't change in OCS.

Normal OCs also had to worry about failing OCS for academic or physical reasons. If that happened, they would be sent out to the fleet as unrated E3 sailors. They would then have to work their way into a specialty and another school while serving the rest of their 3-year enlistment. If I somehow flunked out of OCS, I would return to the NSG as a petty officer 2nd class, since I had earned that rank before OCS. Over the last year, I had decided that CTI2 was the sweet spot in the first four or five years of enlisted service. Your rank got you out of mess duty and other obnoxious chores. Your work experience got you the most interesting job assignments. And, unlike a CTI1 or chief petty officer, you would have only minimal management responsibilities.

OCS also included firefighting training much like that at boot camp. We didn't get rifle range time, even though we spent a lot of time marching around carrying WWI-era Springfield rifles. We did get a chance to qualify with the M1911 Colt Automatic pistol which was the standard military sidearm in the 1970s.

After a few weeks of marching around with a rifle, I found the rifle drill was irritating a few warts on my right hand. I reported to sick bay on a Friday afternoon and a Hospital Corpsman gave me Novocain shots, burned off the warts and bandaged my hand. I returned to my company just in time for the Friday drill and review in the gym. Each company demonstrated their marching and rifle drill skills for the instructors. The company with the best performance got to check out first for weekend liberty. A really bad performance could result in liberty being delayed until after remedial drill on Saturday morning. I nearly put my company in that category. As we were standing with our rifles grounded beside us at parade rest, the numbness in my hand caused me to lose my grip on the barrel of my rifle. Had it fallen to the ground it would surely have marked us for remedial drill! Luckily, the man next to me saw me lose my grip and caught the barrel before the rifle could fall. He pushed it back into my hand and we continued without catching the drill instructor's eye.

After the parade, my company commander, who had caught the motion, asked what happened. I explained the numbness and showed him the bandages under my white glove. He exempted me from drill for two weeks, so I couldn't mess things up for the rest of the company. While exempted from drill and Friday review, I was appointed to be the company clerk. This got me out of Friday reviews for the foreseeable future, as the company clerk had to type up the company liberty list for presentation to the Quarterdeck by the end of the review. OCs who needed remedial instruction or who had lost points in personal or dorm inspections would have restricted liberty. Occasionally, I wouldn't get the list of restricted candidates until late in the afternoon, and I would be typing as fast as I could while the rest of the company was marching in review. On one Friday, I wasn't quite fast enough, and the liberty list was late in arriving at the quarterdeck. I was not a popular person to the candidates waiting to check out. The officer of the deck took the list, looked at my name tag and changed the notation next to my name from 1700 Friday to 1200 Saturday. While my company mates were somewhat mollified by my loss of liberty time, it didn't bother me too much.

Most of my classmates were only peripherally aware that there was more to the Newport Naval Station than the classrooms, dorms, gyms and mess hall of OCS. OCS was my fifth military posting and I had spent some of my free time before the start of the class exploring the base. The base had a good library, a movie theater, and a recreational services photo lab. Over the course of OCS, I probably spent forty or fifty hours in the library and photo lab. I might also have one or two library books alongside my textbooks on my desk. During one room inspection, the senior OC conducting the inspection noted the library books and asked, "Do you really have time to read those?" I replied, "Not much sir.". In reality I usually went through one or two books a week, depending on the amount of time I spent in Newport or the photo lab on weekends. Perhaps it took the inspecting OC more than four or five hours to read a medium-size thriller or detective novel.

The primary recreation for most OCs was to visit the bars at the Officer's club or in downtown Newport. The hot social scene was at the Viking Hotel bar. It had live music and lots of OCs and other Naval personnel mixing, dancing and trying to establish a connection—either with local girls or nurses at the Naval Nursing Corps indoctrination school That school gave the incoming nurses about 4 weeks of schooling on Naval organization and history, how and who to salute and how to wear the uniform. They could do all this in four weeks as, unlike regular OCs, they were already qualified in their profession. The nurses and local girls all seemed to prefer the commissioned officers—who could easily be distinguished from the OCs by uniform difference and rank insignia. Most OCs had very limited civilian wardrobes and generally went to the O club or the Viking in dress blues.

At that time, all the OCs were male. As more women became naval officers other than nurses, later OCS courses included women. I'm not sure how they worked out the logistics of that, since the dorms were like college dorms with bathrooms down the hall.

One of the ceremonial chores rotated amongst the companies was the raising and lowering of the flag in front of the main residence hall. This was usually a very large flag and was raised up a very tall pole. OCS is on a headland on the east shore of Narragansett bay it could get very windy at times. There were several occasions where the flag handlers or line handlers were lifted off their feet by the flag or lines. Flag duty in the winter was particularly hazardous when ice had built up on the pole or lines. Wind and snow also made marching to and from class or to the mess hall very unpleasant. We had wool uniforms and Pea Coats, gloves and good shoes, but your ears were exposed, and you could easily lose your hat. That lead to the universal motto: "When the wind blows, NAVOCS sucks" When back and forth and unobserved by officers, we would sometimes call cadence with variations of, "N A V O C S NAVOCS Really Sucks"

For the last month of OCS, my recruit company commander was appointed Battalion Commander. That is the highest OC rank. The selection is made by the training staff based on academic performance and leadership as a company commander. OC Hudson was a great choice: he was a good company commander, excellent academically and had charisma out the wazoo. He also looked like he was born to the job: about 6'3", athletic build and very handsome. With a little imagination, you could see "Future Admiral" tattooed on the back of his neck. His appointment was fortuitous for me as he got to choose his staff officers. He picked me as Battalion Adjutant. (You can translate "adjutant" as "private secretary"). For the rest of OCS, I typed memos and made up lists for the battalion commander. This exempted me from a lot of drill, sports nights, and duty as a junior watch officer on the quarterdeck. It also meant that I turned in my rifle and checked out a sword that I used at Friday reviews. I marched at the front of the battalion with the other recruit officers.

A milestone in my life occurred at OCS. I bought my first new car. It was a Navy Blue 1971 Chevrolet Vega and cost $2500. I purchased it with a loan from the Navy Federal Credit Union and insured it with USAA. I got the car sometime in February. I didn't drive it too often—just a few trips into Newport with friends and one weekend ski trip to Sugar Loaf in Vermont. It generally sat in the parking lot all week. After one rainy, then freezing week, I opened the door and found about two inches of ice in the driver's side foot well. There was a bad seal around the hood release cable and water had dripped into the car, then frozen. I ran the heater for an hour to melt the ice, then punched a hole in the floor to drain the water. After I gooped up the hood release cable, I had no further leaks. After OCS, I drove the Vega to Pensacola, then across the country to have it shipped to Hawaii from Oakland California.

The Chevy Vega acquired a reputation for poor reliability. It was one of the first cars to use an aluminum engine block with cast iron cylinder sleeves. This reduced the engine weight significantly but there were lots of reports of engine block failures. I never had this problem and the car was generally reliable for the 8 years that I drove it. There were occasional minor problems—but nothing I couldn't fix with parts from a store and hand tools. I liked the Vega so much that I bought another in 1979 when the first started suffering from Hawaiian cancer (body rust caused by salt air exposure). That second Vega got totaled in 1981 when another driver ran a stop sign and struck the rear passenger side of my car. I was spun into a power pole, but not injured. The other driver was OK also—he'd missed the stop sign, but had remembered to fasten his seat belt.

My OCS class had a genuine celebrity as a student. David Eisenhower is the grandson of former president Dwight Eisenhower and was married to Julie Nixon, the daughter of the US President in 1971. David was in another company, so I seldom encountered him at OCS. Julie was living in Newport while David was at OCS, and I would sometimes see them off base. I remember seeing them waiting in line for a movie in Newport. There were a couple of Secret Service agents standing in line with them.

As I recall, David was treated like all other OCs. The only significant difference for the rest of us was that his father-in-law, POTUS, gave our Commissioning Ceremony speech. The week before that ceremony is usually hectic with OCs receiving duty assignments, packing up and practicing for the final drill and parade. It was even more hectic for us, as some changes were required by the Secret Service. The most unusual was that we had to take our drill rifles to the armory to have the firing pins removed. We had never had ammunition for these rifles and weren't even sure they could be fired. However, the Secret Service didn't want a few hundred rifles in the hands of people they hadn't personally checked near the President. Remember, this was less than 10 years after the Kennedy assassination.

An oddity of the commissioning process for former enlisted personnel is that you had to be discharged before re-enlisting to accept your commission. Thus, I have two DD214 discharge papers, one as a CTI2 and the second as a LTJG.

After commissioning and final processing, I left Newport for NSG orientation in Pensacola, Florida. I drove from Newport to Washington DC with two friends who were going to the same course. One was Steve Pennick, the former sonar technician I had met when I first arrived at OCS. The other was a friend from my company whose family lived in Georgia.

We spent about two days in DC. I contacted Joanne, with whom I had traveled in Europe and we went out for dinner. After leaving DC, we dropped off my friend in Georgia and spent a day visiting with his family. He picked up his own car to drive to Pensacola.

THIS IS AN IMPORTANT RECORD
SAFEGUARD IT.

PERSONAL DATA

1. LAST NAME - FIRST NAME - MIDDLE NAME: BORGERSON, Mark John
2. SERVICE NUMBER: B86 28 74
3. SOCIAL SECURITY NUMBER: ███ 68 ███
4. DEPARTMENT, COMPONENT AND BRANCH OR CLASS: NAVY - USNR
5a. GRADE, RATE OR RANK: CTI2 (OCUI2)
b. PAY GRADE: E-5
6. DATE OF RANK: DAY 01 MONTH NOV YEAR 70
7. U.S. CITIZEN: [X] YES [ ] NO
8. PLACE OF BIRTH (City and State or Country): MISSOULA, MONTANA
9. DATE OF BIRTH: DAY ███ MONTH ███ YEAR 46

SELECTIVE SERVICE DATA

10a. SELECTIVE SERVICE NUMBER: 4 4 46 454
b. SELECTIVE SERVICE LOCAL BOARD NUMBER, CITY, COUNTY, STATE AND ZIP CODE: #4 EUREKA, CALIF
c. DATE INDUCTED: DAY 06 MONTH JAN YEAR 69

TRANSFER OR DISCHARGE DATA

11a. TYPE OF TRANSFER OR DISCHARGE: DISCHARGED
b. STATION OR INSTALLATION AT WHICH EFFECTED: NAVAL OFFICER CANDIDATE SCHOOL, NAVAL BASE, NEWPORT, R.I.
c. REASON AND AUTHORITY: TO ACCEPT COMMISSION AS OFFICER IN USNR BUPERSMAN ART 3850220.1b-214
d. EFFECTIVE DATE: DAY 11 MONTH MAR YEAR 71
12. LAST DUTY ASSIGNMENT AND MAJOR COMMAND: SEE ITEM 11b
13a. CHARACTER OF SERVICE: HONORABLE
b. TYPE OF CERTIFICATE ISSUED: DD 256N
14. DISTRICT, AREA COMMAND OR CORPS TO WHICH RESERVIST TRANSFERRED: NOT APPLICABLE
15. REENLISTMENT CODE: RE-2

SERVICE DATA

16. TERMINAL DATE OF RESERVE/UMT & S OBLIGATION: DAY 06 MONTH JAN YEAR 75
17. CURRENT ACTIVE SERVICE OTHER THAN BY INDUCTION
a. SOURCE OF ENTRY: [X] ENLISTED (First Enlistment) [ ] ENLISTED (Prior Service) [ ] REENLISTED [ ] OTHER
b. TERM OF SERVICE (Years): 04
c. DATE OF ENTRY: DAY 07 MONTH JAN YEAR 69
18. PRIOR REGULAR ENLISTMENTS: NONE
19. GRADE, RATE OR RANK AT TIME OF ENTRY INTO CURRENT ACTIVE SVC: SN JC
20. PLACE OF ENTRY INTO CURRENT ACTIVE SERVICE (City and State): EUREKA, CALIF
21. HOME OF RECORD AT TIME OF ENTRY INTO ACTIVE SERVICE (Street, RFD, City, County, State and ZIP Code): ARCATA, HUMBOLDT, CALIF 95521

22. STATEMENT OF SERVICE		YEARS	MONTHS	DAYS
a. CREDITABLE FOR BASIC PAY PURPOSES	(1) NET SERVICE THIS PERIOD	02	02	05
	(2) OTHER SERVICE	00	09	15
	(3) TOTAL (Line (1) plus Line (2))	02	02	05
b. TOTAL ACTIVE SERVICE		02	02	05
c. FOREIGN AND/OR SEA SERVICE		00	09	15

23a. SPECIALTY NUMBER & TITLE: CTI2 2495/0000
b. RELATED CIVILIAN OCCUPATION AND D.O.T. NUMBER: NOT APPLICABLE
24. DECORATIONS, MEDALS, BADGES, COMMENDATIONS, CITATIONS AND CAMPAIGN RIBBONS AWARDED OR AUTHORIZED: NATIONAL DEFENSE SERVICE MEDAL
25. EDUCATION AND TRAINING COMPLETED:
NTC MRPO3&2
NTC CTTRI3&2
DEFLANGINST (FRENCH)
X
X

VA AND EMP. SERVICE DATA

26a. NON-PAY PERIODS/TIME LOST (Preceding Two Years): TL NONE; EXLV NONE
b. DAYS ACCRUED LEAVE PAID: NONE
27a. INSURANCE IN FORCE (NSLI or USGLI): [ ] YES [X] NO
b. AMOUNT OF ALLOTMENT: $ NA
c. MONTH ALLOTMENT DISCONTINUED: NA
28. VA CLAIM NUMBER: C- NA
29. SERVICEMEN'S GROUP LIFE INSURANCE COVERAGE: [ ] $10,000 [ ] $5,000 [ ] NONE X $15,000

REMARKS

30. REMARKS: NOT AVAILABLE FOR SIGNATURE
X
X
X

AUTHENTICATION

31. PERMANENT ADDRESS FOR MAILING PURPOSES AFTER TRANSFER OR DISCHARGE (Street, RFD, City, County, State and ZIP Code): SEE ITEM 21
32. SIGNATURE OF PERSON BEING TRANSFERRED OR DISCHARGED: SEE ITEM 30
33. TYPED NAME, GRADE AND TITLE OF AUTHORIZING OFFICER: I.M.I. MOTZ PERS OFF., BY DIR OF CO
34. SIGNATURE OF OFFICER AUTHORIZED TO SIGN: [signature]

DD FORM 1 JUL 68 214N — PREVIOUS EDITIONS OF THIS FORM ARE OBSOLETE. — ARMED FORCES OF THE UNITED STATES REPORT OF TRANSFER OR DISCHARGE — 3
S/N 0102-002-0200

Figure 10 My First DD-214

This certificate shows that I was promoted to E-5 on 01 November 1970. It took a lot of luck in the timing of exams to get promoted to E-5 in less than two years of active service.

Figure 11 OCS Commissioning Portrait

Chapter 9 Pensacola to Oakland

When we arrived at Pensacola, my two friends and I checked in at the Naval Station and spent a few days in the Bachelor Officers' Quarters while we looked for off-base housing. We found a beach house a short commute from the base. Beach housing was unusually inexpensive, as a hurricane had come through a year before and many houses and yards still showed a lot of damage. The place we rented was concrete-block construction and survived the hurricane, but the lawn, fence, and landscaping had been buried under new sand dunes and scrub grass.

Our course work in Pensacola lasted about three weeks. It was a basic introduction to the Naval Security Group and the US cryptologic community. Much of it was the same material I had learned at Goodfellow Air Force Base as at CTI3. The material was new to my housemates, so we didn't go full animal house during our time in Pensacola, so they could stay awake during classes.

Cross Country Drive

When the orientation course work was done, I packed my car and left for California. I was allowed about two weeks leave and travel time to report to my next station in Hawaii. I had arranged to drive my car to the Oakland Army Base for shipment to Hawaii. I would fly from San Francisco to Honolulu.

I don't remember too much about the cross-country drive. The high point, literally and figuratively, was at Berthoud Pass in Colorado. I discovered that normally-aspirated (carburetor, not computerized fuel injection) engines lose a lot of power above 5000 feet unless the carburetor is adjusted. I was down to about 30 miles per hour near the top of this pass. The downhill part went much more quickly

Near the end of the trip, I stopped in Davis to visit my former roommate. He was then in a graduate program in Animal physiology. He did well enough at that to get accepted into the University of Kansas veterinary school—which is much more difficult than getting into a medical school. The last I checked, he was still a practicing veterinarian in Olathe Kansas.

I also stopped in to visit my brother, Bruce. He was living in Berkeley and taking some courses at UCB. He lived in a basement apartment and was a bit paranoid—his apartment had been broken into a few times by persons probably looking for drugs or easily pawned electronics.

I turned my car in for shipping to Hawaii at the Oakland Army Base, grabbed my luggage and caught a bus to Travis Air Force Base, from where I flew to Hawaii on a DOD-contracted airliner.

Figure 12 High point of Cross-Country drive

Chapter 10 NavCommSta Honolulu

I arrived in Hawaii on April 28, 1971. After my flight landed, I caught a Navy shuttle van to Naval Communications Station, Honolulu—which is located north of Wahiawa, which itself, is about in the middle of the island of Oahu.

I checked in at the quarterdeck and was assigned a room in the Bachelor Officers' Quarters (BOQ). The BOQ at Wahiawa did not meet Navy standards for a number of reasons. The most significant shortcoming was lack of on-base meal service for officers. The Officers' club next to the BOQ only served lunch and happy hour snacks. The best you could do for an evening meal was sandwiches and chicken lumpia and other snacks cooked up by the Filipino stewards. Because on-base housing for junior officers was below standard, we were all allowed to live off base and draw a Basic Allowance for Quarters. This allowance was about $180 in Hawaii at that time. It wasn't enough to rent an apartment near Waikiki but would just about cover a small apartment in a working-class community.

After about a week in the BOQ, my car arrived from California and I moved to a high-rise apartment in near downtown Honolulu. I shared the $800 monthly rent with 4 other officers. It was a nice apartment and close to the night-life, but it meant sharing a bedroom and commuting about 30 minutes each way to and from Wahiawa. That arrangement only lasted about three weeks. At the end of the month, I found a two-bedroom apartment on Lakeview Circle between Wahiawa and Schofield Barracks.

I shared that apartment with Mike B another Ensign stationed in Wahiawa. The apartments were single-story duplexes with single-wall construction (no sheet rock inside the wall studs). The complex dated back to WWII days and had once housed civilian workers at Schofield Barracks. In 1971, it housed a mixture of civilians and military personnel. I think the rent was about $150 per month, so Mike and I had plenty of BAQ left for groceries and utilities.

For most of the spring of 1971, I worked as an assistant division officer in the Naval Security Group department at Wahiawa. I was in the W-47 division, which was the general operations SIGINT division. We had a about 20 linguists and Morse code operators listening for both Soviet military signals and signals related to the south Pacific. We intercepted high-frequency radio teletype, Morse code, and voice signals between the main site and outlying bases and weather stations. Our predictions weren't perfect, but were adequate to warn the Air Force so that they could dispatch surveillance flights from Hickam Air Force Base in Honolulu

My days in W-47 were spent managing the linguists, Morse, and TTY operators, reviewing the intercepted traffic and doing some translations. There wasn't enough work to keep me busy all the time, so I started on a basic code-breaking correspondence course. Also had the facilities manager set up a voice receiving station in the basement where I could listen in on voice signals Most of the traffic I recorded was routine discussions of personnel and logistics matters. It had no great intelligence value but did give me some practice at listening to and translating

During one of my morning listening sessions,

The other division in our building was W-48, the High Frequency Direction Finding (HFDF) division. We worked in the concrete building inside the Circularly Disposed Antenna Array (CDAA) or Wullenweber antenna system. This antenna system could receive signals from all directions and measure the bearing to the signal with an accuracy of better than one degree. It was designed and built to receive signals from Soviet submarines. There was a network of about nine of these CDAAs around the Pacific basin. The HFDF stations were connected by teletype to a master station at NAVCOMMSTA Honolulu. A group of receiver operators in W48 would listen for the Soviet submarine signals, and when one was heard, they would notify the master station. Other stations in the network would search for the same signal, and if it was heard, send the bearing to the master station. If enough stations got good bearings, they could triangulate the position of the emitter.

A, at the time highly-classified, program called Classic Bullseye involved continuously recording large parts of the HF spectrum on high-speed tape recorders. The recorded combination of radio signals and antenna pointing data allowed what was called Wideband HFDF or retrospective HFDF. The Soviet [redacted] had started to use short compressed burst communications to report to their bases. [redacted]

[redacted] Appendix A has a great overview of the SIGINT activities at Wahiawa.

Classic Bullseye and the associated submarine tracking effort, called PELAGIC, were tightly controlled compartmented programs. I didn't have access to these programs when I worked in W-47 but found out something about them when I was later given access to several highly sensitive programs, so I could stand watches at NAVSECGRU Pacific headquarters at Pearl Harbor. I also got a lot more access later, when I was a division officer for a Special Intelligence Communications relay center. At the time,

I was in W-47, I just knew the people in the wideband room spent a lot of time changing tapes and working with HFDF receivers.

In addition to my W-47 duties, I was assigned to temporary duty as a division officer for a program called Pony Express. This was a program that monitored telemetry signals from Soviet missile tests in the Pacific region. I spent several weeks training with enlisted personnel aboard specially-configured destroyer escorts at Pearl Harbor. On July 11, 1971, it appeared that a missile test was imminent, and I was flown to Yokusuka Japan with 10 enlisted personnel to work the telemetry collection effort aboard the USS Claud Jones (DE-1033). We set up our equipment and stowed our classified material and waited for further developments. During some free time, I went ashore and did some shopping---a Canon lens for myself and cultured pearls for my mom and sister. After about four days, NSA realized it was a false alarm and the mission was canceled. Along with my enlisted contingent, I got on a high-speed train to Narita airport to return to Hawaii. We were scheduled to return on a new Pan AM wide-body jet, the Boeing 747. When we arrived at the airport, we found the flight had been delayed a day due to a mechanical issue. There weren't enough available seats on other flights, and our classified materials had to fly on an American airline, so we were sent to the Imperial hotel in Tokyo where we would spend the night before boarding the flight the next day. Some of the enlisted men went out drinking and carousing that evening. I was content to go to the fancy restaurant in the hotel and order a steak dinner and bottle of wine. The bill exceeded my $35 food allowance, but Pan Am never tried to recover the extra $15.

The next morning, I boarded a van for the airport with my sailors, a couple of whom were well under the weather from their evening activities. At the airport, I took a couple of sailors to a secure military storage facility where we picked up our footlockers of classified TEXTA, TECHINS, and key cards. When we were ready to board, I had to accompany the footlockers out to the ramp to make sure they were stowed in the baggage hold just before it was closed. I then boarded the flight and we took off. Because I was out on the ramp watching the footlockers, I didn't do a head count as we boarded, and I hadn't properly delegated that duty. About two hours out of Japan, one of my sailors found me and said, "Mister Borgerson, we seem to be missing CTO2 M." We found a flight attendant and verified that he was not on the plane. I had visions of being court martialed for losing a highly-trained specialist in Narita airport. We contacted the aircraft radio operator and he called PAN AM Tokyo operations. It turned out that, while several sailors had laid out for recuperative naps in the airport, CTO2 M had awakened about half an hour after the plane left. He acted promptly and effectively by contacting the military travel office at Narita and got booked on the next flight out to Honolulu.

When we arrived at Honolulu, the Pan Am gate attendant notified us that M would be arriving at a different gate in about 45 minutes. By the time we got our bags and loaded our footlockers in the Navy van, I had just enough time to greet M at the gate. I got a sheepish "Sorry Sir" and he collected his sea bag and we took the van to Pearl Harbor. In my travel summary, I noted M's changed flight as due to "classified materials transport problems". It was true for some values of 'true', and I figured the travel office wouldn't ask too many questions as they didn't like to get involved with classified materials as it engendered lots of extra paperwork. Since there was no extra cost or delay in operations, there were no repercussions---except the hazing M and I received from the rest of the team.

In August of 1971, testing was finished for the year and I was released from my duties in W-47 and assigned as the division officer for the Special Intelligence Communications (SPINTCOMM) relay center at the Kunia. The Kunia facility had been built as secure armaments storage during World War II. It was a 3-story concrete building that had been built in a ravine south of Schofield Barracks, then covered with about 60 feet of dirt. Pineapples were planted and harvested over the top of the building. You entered the facility by walking down a 1/8 mile of tunnel to the building.

The SPINTCOMM relay center received highly classified teletype traffic to and from Far East Commands and government facilities from Vietnam to Korea. The traffic was relayed to recipients in Hawaii and on the mainland, such as CIA, DIA, NSA and cryptologic agencies, the White House, etc. The messages were nearly all classified and many were from very tightly-controlled compartmented programs. At that time, teletype messages came in and were decoded and transferred to paper tape by the CTO communicators. The operators would review the decoded messages and attempt to correct any garbles due to problems in transmission or decryption. This meant that highly classified traffic for special programs that might have only a very small access list could be read by any of about a hundred enlisted communicators. As a result, we all had to have access to these programs. It was at this time I was introduced to programs such as Talent Keyhole (Satellite and SR-71 programs), Holystone BYEMAN, (special SIGINT, often from satellites), PELAGIC (HFDF tracking), Delta Series COMINT (COMINT with very restricted access).

When you reported to a new station, the NSG would send a classified message specifying your clearance level and access. When I reported to NAVCOMMSTA Honolulu, my message was something like this:

BORGERSON, M. ENS. TS/CRYPTO/SI Top Secret, Cryptographic, Special Intelligence. This was the basic clearance and access level for all NSG personnel.

When I was at Kunia, my message looked like:

BORGERSON, M. ENS. TS/CRYPTO/SI TK/TK

The access digraphs were repeated to reduce the probability that a message garble would give someone improper access.

At that time, the military was just starting to transition to computer-controlled message relays. There was little confidence that these relays were reliable and secure enough to handle very sensitive traffic, so such traffic was still handled by enlisted personnel on twenty-four-hour duty at relay centers.

My division at Kunia was really run by three very senior chief petty officers. I was there primarily to sign for classified material and to write personnel evaluations of the chiefs. Myself, the chiefs, and about 10 other office workers and maintenance technicians were day workers: Monday through Friday 8AM to 5PM. The rest of the CTs were organized into four watch sections for 24-hour operations. There wasn't too much to do each day, so I would wander the floor reading message traffic, work on my cryptology correspondence course, and receive cribbage and Acey-Deucy lessons from the chiefs. We could hear the clatter of the teletypes, the chatter of the operators and the occasional cry: "SPO on DIRC ALPHA". This meant that Special Project Operations traffic had been received on circuit DIRC Alpha. This traffic had to be reviewed logged by the senior petty officer on watch.

Amongst the circuits we handled were what was known as "Admiral to Admiral" circuits. These were circuits where admirals and generals could send informal, back-channel messages to their peers. Most of the messages concerned travel planning and strategy discussions. However, we would sometimes forward Top-Secret messages arranging golfing tee times.

Early in 1972, SPINTCOMM forwarded a general-to-admiral message from the Director of the Defense Nuclear Agency to CINCPACFLT about setting up an NSG detachment on a ship to be assigned to cover the [redacted] [redacted] One of the senior petty officers knew that I was a French linguist and had worked on [redacted] at W-47. He showed me the message on the theory that I might be interested. I was. However, I didn't know quite what to do about it. I wanted to volunteer to be the NSG detachment division officer, but I was reluctant to reveal that I knew about the position because I had read an admiral-to-admiral message. I solved the problem by contacting the W-47 division officer and reminding him that I would like to return to W-47 if they needed a [redacted] division officer that summer. My timing might have seemed suspicious to the W-47 division officer when the orders to set up a direct support detachment filtered down to him a few weeks later. Later in the spring, I was assigned to be the NSG division officer on board the USNS Wheeling. That assignment was repeated in 1973 and is the subject of another chapter in this story,

In the fall of 1971, several of my DLI classmates had finished their two-year "hardship" tour in Morocco and were reassigned to NAVCOMMSTA Honolulu. I was reunited with Stephen H, Dan H and Jim G from my DLI class. There was a bit of an officer-enlisted divide, but we did manage to exchange some stories about our time since I had left for OCS. We got a lot more time to talk later when we were aboard the USNS Wheeling. At about the same time CTM3 Mark B, with whom I'd visited the Oktoberfest, arrived in Hawaii. Mark and I would go on to spend a lot of time working in the Schofield Barracks theater group.

Getting to the SPINTCOMM facility at Kunia was a matter of walking from the parking lot to the gate at the outside of the tunnel, showing the guard your pass and walking about 250 yards down the tunnel to an elevator that would take you to your floor. This wasn't an issue except for a few weeks at the end of December. One Sunday Mark and I went out to Waimea Bay to try some body surfing in about a six-foot shore break. We got a couple of good rides---perhaps almost as good as some of the local six-year-olds. Near the end of the day I was popping up to check the swell before a ride and pushed up off the bottom. I felt a sharp pain in the second toe of my right foot. I had caught the toe on a rock embedded in the sand. I limped ashore and the toe was starting to swell up badly. Mark and I got back to the NAVCOMMSTA and I went to sick bay. I had a green-stick fracture of the toe bone. The corpsman said it wasn't serious and taped it to the big toe. He issued crutches and said to use them until the swelling went down and I could walk comfortably. For the next two weeks, I walked around with crutches and white tennis shoes. My normal black shoes wouldn't fit on the injured foot. At a department meeting an officer known for being picky about uniform standards asked why I was out of uniform. I pointed to the crutches leaning in the corner and said, "I thought two white shoes would look better than one white and one black." Walking down the tunnel on crutches wasn't fun, but it did provide a good incentive to heal up and get back to normal shoes.

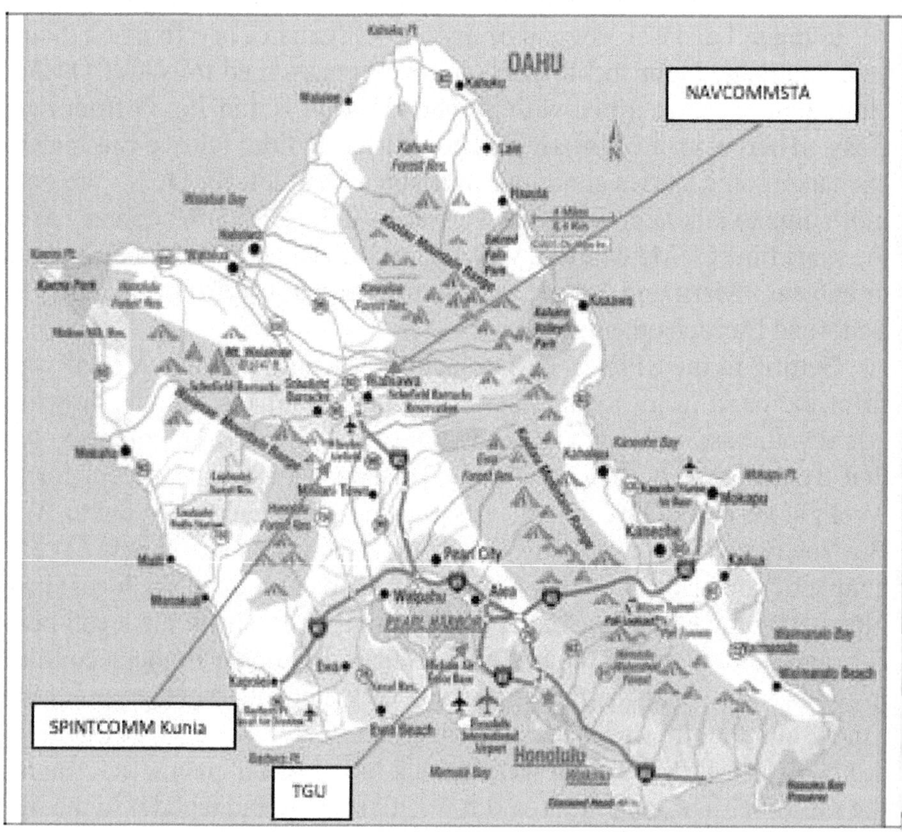

Figure 13 The places I worked on Oahu

Figure 14 FRD-10 Antenna at NavCommSta Honolulu
The Ops building with W-47 and W-48 is in the center of the antenna.

Chapter 11 RPIO and TGU Pearl Harbor

In early January of 1972, I was released as the SPINTCOMM division officer and another junior officer took over that duty. Along with three other junior officers, I was assigned to take a course in the procedures for issuing registered materials at the Pearl Harbor Registered Publications Issuing Office (RPIO). Registered publications included classified manuals and cryptographic keying materials, as well as the actual cryptographic equipment. There were very strict accounting procedures for this material to prevent inadvertent disclosure of classified information. Most of the work at the RPIO was done by civilians with appropriate security clearances, but regulations required the signature of a commissioned officer for each transfer. The course included a lot of accounting and paperwork detail and we all found it to be rather boring. While our instructors stressed the importance of proper accountability for classified materials, RPIO duty was widely considered to be a dead-end job. It had no particular relevance to the main career advancement paths in the Naval Security Group.

While in this class at Pearl Harbor, I found that the commute from my apartment in Wahiawa to the RPIO was hot and tedious. To avoid the commute, I moved into the BOQ at Pearl Harbor. This BOQ was a far cry from the one at Wahiawa. There was a nearby officers' mess and the rooms had a small kitchenette. There were several O-clubs on base with full meal service. It was comfortable, but I only spent about a month there.

About a month after I finished the RPIO training I got TAD orders to report to the Technical Guidance Unit (TGU) at Pearl Harbor. TGU was an organization that reported directly to Commander, Naval Security Group Pacific. It consisted of about 6 officers and 80 enlisted CTs who provided specialized technicians for temporary assignments aboard ships and submarines. The TGU was divided between surface support technicians who served aboard the Pony Express Destroyer Escorts, and submarine support technicians ███████████████████████████████

█. The submarine operations, under the code name HOLYSTONE, were tightly compartmented and many of the CTs and officers in surface direct support did not have access to the program. One side effect of that compartmentation was that the submarine support personnel had a separate training area off limits to the surface sailors. There was also a friendly rivalry between the surface and submarine personnel that got played out in softball games and other activities.

The TGU was housed in building 397, a two-story poured concrete building in the middle of the Pearl Harbor Naval Shipyard. The bottom floor housed a telephone exchange—as it had since WWII. The upper floor contained administrative offices and training spaces for the CTs, which we reached via an external staircase.

There was one shared equipment space, which contained a WLR-1 broadband electronic intelligence receiver. This piece of equipment was notoriously unreliable and had received the nickname "Squirrely Whirly One". It was new to me and I spent a lot of time playing around with it when it was not in use for training. Between Pearl Harbor and the adjacent Hickam Air Force Base, there were lots of radars as well as shipboard and aircraft VHF and UHF signals to practice with. I remember working on that receiver one day when President Ford visited Hawaii. I managed to copy some Secret Service communications and part of a mobile telephone call from a member of the President's staff. I erased the tapes immediately as NSA regulations forbade the intentional interception of communications between US citizens.

One of the collateral duties I was assigned while at TGU was Junior Officer of the Watch (JOOW) at the NAG CincPacFlt headquarters. The duty of the JOOW was basically to sit around and drink coffee or sodas during the eve and mid watches and to call senior officers on an alert list if anything exciting happened. Officers who stood these watches had to be cleared for all the sensitive compartmented programs, as there was no telling what type of crisis might occur. Only myself and one other TGU Pony Express officer had the appropriate clearances—in my case as a result of my assignment to the SPINTCOMM relay center. I only stood about three watches in about 18 months. CincPacFlt didn't like to assign the Pony Express riders to a regular watch rotation because there was too much likelihood that the officer could be deployed when they were scheduled for a watch. As a result, I was assigned a watch only when the normal watchstanders were on leave.

NSG CincPacFlt had a low-tech op-center room where the staff O-branchers would send messages needing immediate decisions. There were about 4 O-branchers and a chief on duty there. Nothing unusual happened on my watches and I never had to wake up a Captain or Admiral in the middle of the night.

A secondary duty of the watchstanders was to walk through the staff offices and check for security breaches—classified documents left out on desks, blinds not lowered, etc. etc. This was the most interesting part of the watch for me one night. As I was wandering through the offices, I check found an unlocked desk drawer containing a highly classified report on HFDF results. That got me to wondering what other sloppy security practices were in play. I went over to a row of combination-lock file cabinets and started looking for unlocked cabinets. I didn't find any. Then I thought to check for lazy combinations: 0 0-0-0 (the default for new cabinets), 1-2-3-4, etc. The third cabinet I tried opened with 0-0-0-0! The cabinet was full of secret and top-secret files on HFDF tracking of Soviet submarines and US submarine reconnaissance operations. I took the files from the desk drawer back to the op center where the chief locked them up in a safe. I made a note in the log about the desk drawer and the 0-0-0-0 combination. I suspect that those log entries proved rather embarrassing to one or more staff officers. I was never called to testify at a formal investigation, so I guess nobody thought there was any real risk of compromise of the classified information—just sloppy security procedures that needed to be cleaned up.

Chapter 12 Unclassified Life

When you work in a secure area and everything you do is classified, you develop an internal firewall that separates work from the rest of your life. You don't talk shop outside the shop. That causes a lot of stress for personnel that are married or in serious relationships. As a result of that stress, alcoholism and divorce were common problems in the Naval Security Group. When I was an enlisted CT and early in my years as an officer, this was not much of a problem. I was working and relaxing with other CTs or SecGru officers and we all maintained the separation between our classified and unclassified lives. After I had been at Wahiawa for a few months, I started working evenings and weekends with the little theater group sponsored by the Army recreational services at Schofield Barracks.

Late in the summer of 1971, I received a letter from my college girlfriend, Rebecka (Becka) Q. She had graduated from UC Davis a year after me, gotten her teaching credential, and was teaching at a high school south of Sacramento. I had last seen her just before I left for Morocco about a year before. We had kept in touch with occasional letters. This summer she, and another of the female teachers at her school had saved enough money for a three-week trip to Malaysia and Hawaii. She and her friend would be stopping on Oahu on their return and wondered if I could put them up for a day or two and accompany them on a trip to the island of Hawaii. There was a bit of an ulterior motive in the invitation—it was difficult for people under 25 years of age to rent a car. I was just under 25, but military officers were considered more responsible and could rent cars from most companies.

I applied for and was granted a week of leave to travel to the Big Island. I picked up Becka and her friend at the airport and we returned to my apartment in Wahiawa. A complicating factor was that Mike and I had experienced a sewer backup and overflow just a few days before. We drained the overflow by drilling holes in the floor and cleaned up as best we could, but there was still a nasty odor in the apartment. I explained this to

Becka and she said it didn't smell much worse than some of the places they had stayed in Malaysia, but it was good that we were leaving for Hawaii the next day. We spent about 5 days touring around the Big Island, camping in parks and on beaches. I had my own tent and minimal camping gear. Becka and her friend were better equipped as they had been camping much of their trip.

At the end of the trip, we flew back to Honolulu and Becka and her friend flew back to San Francisco. The trip was pretty much the end of my relationship with Becka. We agreed that we were pursuing different lives and different places and there was little probability of a future closer relationship. I picked up my car at long-term parking and returned to life in W-47 at Wahiawa.

Commuting between Pearl Harbor and Schofield barracks, where I worked with the US Army Hawaii theater group, started to get on my nerves, so I moved to a one-bedroom apartment in Waipahu, which is about half way between Pearl Harbor and Schofield Barracks and Wahiawa, where my girlfriend lived.

Working with the theater group provided both opportunities and responsibilities. I worked with and got to know very well several soldiers from Schofield Barracks, some Air Force personnel from Wheeler Air Force Base, and civilians from the Wahiawa area. For the most part, these people had unclassified jobs and could talk about the daily challenges and frustrations of their jobs. After a few initial queries, the theater people soon learned that I wouldn't talk about work beyond saying that it involved communications and was classified. This wasn't all that unusual as some of the other people had some classified aspects to their jobs or some high-stress or combat operations which were off-limits for casual discussion

There were a few other things that complicated my life with the theater group. The first was that I was the only officer working in the backstage crew on a regular basis. There were some officers and officer's wives that appeared in the cast of some plays, but the backstage crew was uncertain how to treat an officer that would wire a light or paint a flat. At first, this caused a bit of friction. I soon made sure to arrive out of uniform and had everyone calling me "Mark" and not "Mr. Borgerson"

A second factor that complicated life in the group was that I was almost always on 24-hour call to go to sea for Pony Express operations. I told the group director about this limitation and that I would probably not try out for any major speaking roles for which they could not have a ready understudy. Working back stage made more sense, as there were usually several replacements for stage manager, construction or lighting duties.

Soon after I started working with the theater group, I ran into Mark B, my buddy from Morocco. He had been transferred to Hawaii earlier in the year. I recruited him into the theater group as a stage hand and construction worker. Soon after we started working there we ran into a problem. Someone would yell out "Mark, can you give me a hand with this cable?" We would yell back "Which Mark?". After this happened a few times, I pulled seniority and rank and decided I was Mark 1 and he was Mark 2. We announced the new naming conventions and the confusion was greatly diminished.

Mark 2 and I spent a lot of time snorkeling and body surfing at Waimea bay on weekends. As a maintenance technician, he worked a normal day shift with evenings weekends off. Besides being a good friend with a lot of common interests, Mark 2 was very outgoing and never hesitated to talk to the pretty girls at the beach. One afternoon in December of 1971, he introduced himself to a couple of girls and started to tell tales of his explorations of Morocco and Europe. He called me over to back up his stories—which were mostly true. I started to talk to one of the girls, Fran, a very pretty blonde with a big floppy hat and a swimsuit that showed a LOT of cleavage. After a while I asked if I could take her picture and she consented. I found out that she was a college senior and Navy dependent and her father was stationed at Pearl Harbor. A little later, I had asked her out for the next weekend and promised that I would develop the pictures and bring her some prints. She gave me her family address in the Makalapa Heights housing area of Pearl Harbor. That should have been a warning sign—only very senior officers live in that area.

I developed the film and made up two or three black and white 8x10 prints. The photos turned out very well. She had a cute smile and the floppy hat was a great accessory that shaded her face enough to let her smile without having to squint in the sunlight. One of the photos did show a lot of cleavage, but I figured she knew what she was showing and would like the picture.

The next Saturday I arrived at her father's house and knocked on the door, envelope of prints in hand. Fran answered the door and invited me in.

I showed her the prints and she thought they were very good. She ran into another room to show them to her father. They returned in a few minutes and he was in uniform—a full captain with chaplain's insignia. I had just given a set of very sexy pictures to the daughter of the senior chaplain of the Pearl Harbor Naval Station! He was very gracious about the photos but said that Fran might have to tone down her choice of swimsuits in the future. We went out to dinner at a restaurant in Waikiki that night. We talked about the theater group and she said that she would try to get to one of the plays. We had one other date for drinks and dancing at the Pearl Harbor Officers' club. She went back to the mainland to college and we never met again. Shortly after that I started dating Anne Finnigan, one of the prop and costume workers in the theater group.

Here are the plays I worked on at Schofield, in chronological order:

Here's Love December 1971. This play is a musical version of the Christmas story "Miracle on 34th Street". I did set work and played a bit part as the bailiff in the courtroom trial of Kris Kringle.

After the Fall January 1972. This was a somber Arthur Miller drama. I contributed to the set design and construction and was the stage manager. The set was a series of ramps and platforms and received the following review in the Honolulu Star-Bulletin: "On a barren, multilevel set, Quentin's (Miller's) story unfolds, jiggling back and forth in time with the use of extremely precise puddles of light that work fairly well in the Schofield Production" . That was about the best part of the review.

How Now Dow Jones April 1972. This is a musical comedy about love on Wall Street. I did a lot of the set design and construction for this play. The set multiple levels with stairways and railings. I had to borrow a welding set from the Schofield motor pool to construct the frames for the 8-foot-high platforms at the right and left of the stage. Robin and I did a lot of stomping and bouncing on the platforms to make sure the actors would be safe and could move around without a lot of creaks from the structure.

Sammy Davis USO Show Fall 1972. Sammy Davis Jr. stopped at Schofield Barracks to do a show on his return from a tour in Vietnam. He performed in the large amphitheater at Schofield and Robin Moore and I were picked to run the follow spots for the show. We were a bit nervous working with a big-time star, but managed to get most of the cues right and the show as a big success.

Born Yesterday March 1973. I was stage manager and did set construction for this comedy about a junk dealer who is trying to bribe a congressman and his relationship with his showgirl mistress.

Celebration May 1973. I did set design and construction for this avant-garde musical.

Bus Stop March 1974. I did set design and construction for this play about a cowboy and a would-be singer who meet in a diner when a bus is delayed by a snow storm. I was familiar with the play as I had done set work for it while in Morocco.

Company April 1974. I did general backstage work and lighting on this musical. It was the last play I worked on in Hawaii. I actually stayed on in Hawaii for several weeks after I was released from active duty at the end of March, so I could work on the play.

I had a major backstage role in 7 plays in about 2-1/2 years. I also did some minor work on a few others but missed out on at least six others while I was away for the summers of 1972 and 1973. Just before I left Hawaii, I got a certificate of appreciation from the US Army Hawaii.

The Schofield theater group was the center of my social life for most of the time I was in Hawaii. I met the two women I dated, Anne F and Janet R, at the theater. Both these women were elementary school teachers in parochial schools in Wahiawa and were best friends. I started dating Anne in early 1972 and we continued that relationship until the next Christmas. During the Christmas break, Anne flew back to Boston to visit her family. During that time Janet and I went out a few times. Early in 1973, Janet and I had grown close enough that I had to discuss the issue with Anne. We broke up and I continued to see Janet. This caused a major social upheaval in the theater group as the two women were on much less friendly terms for several months. Later in 1973, Anne started dating an Army officer and the two women reached a truce. Anne married the Army officer late in 1973. Needless to say, I was not invited to the wedding—but Janet was.

A major element in theater group social life was the cast party at the end of each theater production. These parties sometimes occurred at the home of Grace P at Schofield. She was a frequent backstage worker and actress for the theater group. Her husband was a colonel in the Army and their quarters were large enough to host the full cast and crew.

Before one cast party at the Grace P's I came close to getting a DUI. I had downed a few beers with Robin Moore as we closed down the theater. I drove him to his quarters at Wheeler AFB, so he could change clothes. We then headed back to Schofield to go to the party at the Pickett's. As I turned into the gate, I misjudged the turn and started into the exit lane. I stopped before the gate, backed up and pulled into the entrance lane. The gate sentry gave me the stink eye while saluting the officer sticker on my car and asked "Sir, do you know where you are going?" I had visions of MPs and stockades and reports to my CO, so I did a bit of shameless name dropping and said, "I'm going to a party at Colonel P's quarters". The gate guards knew about the party, as several civilians unfamiliar with the base housing area would be attending. The guard called for a guide jeep to get me to the party. I'm sure he was also observing my driving skills. After I reached the party, I found Grace and told her about my poorly timed entrance. She laughed and said "We told the MPs we'd keep track of the drinking and driving issues at the end of the party. I didn't think we'd have to worry about the start of the party." I made sure to leave the party sober and via a different gate.

The SECGRU officers with whom I worked also had some parties, but many of them were married and the parties were usually pretty tame. I went to one party at the home of Marty a single officer who lived on the beach in Haleiwa.

Marty had a Sunfish sailboat that he sailed in the waters in front of his house. Later in the party, we took that boat out just beyond the reef and were sailing along nicely when a small hump back whale surface and blew about 20 yards away. Marty yelled "Let's get a closer look!" I replied "He's a lot bigger and heavier than us. If you want a closer look, I'll swim ashore, captain Ahab!". We let that one get away, but it made for a lot of future sea stories.

One notable party at Wahiawa was thrown by the assistant NSG department head at the NAVCOMMSTA. LCDR David Lewis had a greeting party for newly-arrived officers about twice a year. The parties were known for his sea stories and his homemade mango brandy. The brandy was made by freezing mango wine and filtering out the ice crystals to leave a pretty strong brew. The sea stories were also interesting as LCDR Lewis had been aboard the USS Liberty when it was strafed and torpedoed by the Israelis during the 1967 Arab-Israeli war. He had been wounded in the attack and received a reparation payment from the Israeli government. He was certain that the Israelis had known they were attacking an American ship.

I did a quite a few day sailing trips with Anne, Janet, and other friends from the theater group. The Navy had a yacht club at Pearl Harbor, but it was for boat owners and provided no opportunity for casual sailing. However, the recreational services as Hickam AFB had a marina that rented sailboats to military personnel who could pass a basic boating test. I passed the test and would reserve a Rhodes 19-foot Daysailer for a day. I would invite four of five friends and we would pack a picnic lunch and spend the day sailing around the entrance to Honolulu harbor. The sailing area was just under the landing path to Hickam/Honolulu International. Landing planes would go overhead only about 100 feet up. Being that close to a landing 747 or C-5 Galaxy was quite impressive. We would sometimes sail outside the harbor entrance to a point where the water was several hundred feet deep. It was sort of spooky to look down into the water and see your shadow disappear into infinity. We only rarely jumped off the boat to swim out there---it was just too unnerving to think of all that water underneath us.

Figure 15 Rehearsal photo from After the Fall

Figure 16 Rehearsal photo from Born Yesterday

Chapter 13 USNS Wheeling and Hula Hoop

Until this point, I have recounted events pretty much in chronological order. I am going to depart from that for the next two chapters. My two assignments to the USNS Wheeling for Hula Hoop operations occurred during 1972 and 1973. I spent about three months each summer aboard the USNS Wheeling (TAGM-8) providing communications support to projects observing French nuclear testing in the South Pacific. There were a lot of code names associated with these projects, but I am going to generalize and refer to the missions under the code name Hula Hoop. These two missions had a lot in common and are grouped into this chapter.

I have also grouped about half a dozen Pony Express missions on the USS McMorris and USS Claud Jones into a single chapter. These missions also all had a lot in common, though they were spread out over about two years. If I kept things in strict chronological order, the reader might suffer whiplash as I switched back and forth between Pony Express, shore duty at TGU, and Hula Hoop missions. Appendix B contains excerpts from several declassified documents about the US observation of French nuclear testing.

The goal of Hula Hoop was to collect data on the atmospheric nuclear tests being conducted by the French at Mururoa Atoll. As a signatory to the Limited Test Ban Treaty (LTBT), the US was not allowed to conduct atmospheric nuclear tests. However, as part of the push to develop anti-missile defenses, the US needed to know how effective missile control radars would be if there were nuclear detonations near the missiles or radars. These tests could not be done with underground detonations, so the US had to test radars with the French atmospheric tests. China was not an LTBT signatory, but its tests were in the Eastern interior of China at Lop Nur, and were inaccessible to US radar testing.

The Wheeling was a missile range tracking ship and had large parabolic dish antennas for both tracking radars and telemetry reception. Additional radars of the type used in missile guidance were mounted in vans on the aft helicopter deck of the Wheeling. The Wheeling started life in 1944 as the

Seton Hall Victory, a cargo ship built at Oregon Shipbuilding in May of 1945. After several other roles, it was converted to a missile range tracking ship in 1962 and assigned to the Navy's Pacific Missile Testing Range at Port Hueneme, California. Port Hueneme is near Oxnard and just north of Santa Barbara on the southern California coast.

In addition to the Wheeling, the Air Force would provide two specially-configured RC-135 reconnaissance aircraft to conduct additional tests. These aircraft were also used to monitor the 1971 French tests. When I worked at NAVCOMMSTA HONO in 1971, we had provided communications intelligence support for those missions. However, it was very difficult to determine the time of a test with the precision the Air Force needed from only the HF radio signals available in Wahiawa.

As the RC-135 aircraft could only stay on station for a few hours, it was imperative that they have good advance information on the timing of the tests. The addition of an NSG direct support unit aboard the Wheeling was intended to provide more detailed information than was possible with only HF intercepts from Wahiawa.

Shortly after I was assigned to the Wheeling for Hula Hoop in May of 1972, I flew to Los Angeles, then Oxnard California and reported aboard the Wheeling, which was docked at nearby Port Hueneme. I was accompanied by several CTM maintenance technicians and other CTs who would handle the installation of our equipment aboard the Wheeling. When I arrived at the ship, I started my time on board with a major faux pas. I had been ordered to report to the Hula Hoop military officer in charge, commander Richard B. I reported to him and went below to the forward chartroom (under the forward helipad in the photo) to supervise the equipment installation. About half an hour later, an announcement came over the ship's PA system, "Ensign Borgerson report to the bridge."

I climbed up to the bridge and was told by commander B that I needed to talk to the captain. I crossed the bridge and stood at attention in front of the ship's captain, E.R. Gibson. Captain Gibson was a formidable presence: over 6 feet tall with silver hair and beard. He did not look happy.

"Ensign Borgerson, is it not the naval custom that an officer reporting aboard a ship present himself to the ship's captain?"

Oops! I had reported to commander B because he was listed on my orders and was cleared for signal intelligence information. I had not though to report to the captain, who was a civilian-licensed ship's master and was not cleared for SIGINT. Captain Gibson was a bit upset about my violation of protocol and let me know it. After a minor chewing-out, he welcomed me aboard. He acknowledged that he wasn't cleared for the details of my duties, but that I shouldn't feel unique. He said there were a lot of spooky types aboard and he would rely on commander B to keep him informed of the things he needed to know to put the Wheeling in the right place at the right time.

The SECGRU detachment aboard the Wheeling consisted of about 8 people:

One of the linguists was usually a CTI1 and served as the leading petty officer for the division. He set up the watch schedules made sure a minimum of military decorum—which was difficult on a ship where 85 percent of the crew were civilians.

We set up our equipment in the forward chartroom, which was just forward of the ship's radio room. We had to install our receivers and recorders in equipment racks, run cables to the radio room for our incoming and outgoing communications, and run cables to antennas mounted on deck. We also had several teletype machines, both for printing communications. We also had our own crypto gear for encrypting our communications signals to and from the radio room. The forward chartroom initially contained only several large cases of spare hydrographic charts, which we moved to the front of the room. There was plenty of room for both our equipment and the chart cases, as the room was about 16 by 30 feet. There was a hatch at the front of the room leading to a storage compartment with an escape hatch to the forward deck. We used that room to store our equipment cases and bags of shredded paper generated when we destroyed classified teletype messages.

In addition to our racks of [illegible] equipment along the port side of the chartroom, we had several tables and desks with typewriters, a transcription station with a tape recorder and typewriter, and our communications equipment. At the aft end of the room was a curtained doorway to the passage to the radio room and the rest of the ship. There was an electronic lock on the door. Other than myself and the CTs, only commander B, two of the senior scientists, and one Air Force officer from the Defense Nuclear Agency, had the combination to the electronic lock.

Working conditions aboard the Wheeling were heavenly when compared to the conditions aboard the Pony Express DEs. As part of the conversion to a missile range tracking ship, the lower six to eight feet of the Wheeling's cargo holds had been filled with concrete ballast. The ballast and the larger, rectangular hull cross-section made the Wheeling a very stable ship. The stability was necessary for the proper operation of the large tracking antennas. It was a great fringe benefit for the people aboard. We could put down coffee and sodas without worrying about them sliding off the desk or table. If the weather was calm, you could hardly tell the ship was at sea.

The accommodations aboard the Wheeling were also much better than most naval ships. The enlisted CTs were in two or four-man cabins with two-high bunk beds—quite an improvement over the 30-man berthing compartment and 3-high pipe and canvas racks on the DEs. Military officers and senior scientist had one or two-man cabins with a single set of bunk beds, a desk and lockers. Officers also had a steward to clean the cabin, make up the bunks, change the linen and handle your laundry.

There was a separate wardroom/mess room for the military officers. We had our own dedicated steward, Roy, to serve our food from the general mess. We would make our choices from a menu that generally took up about a full single-spaced page for each day's three meals. Roy would take our orders and return with plates from the galley. The wardroom also had a cupboard for snacks, a coffee maker, and a small refrigerator with milk and juice. We could show up for meals at any time during about a two-hour interval: 6 to 8AM, 11AM to 1PM and 5 to 6:30PM. The 5 to 6 officers that ate in our mess soon settled on a mutually agreeable time and usually appeared within about 15 minutes of the middle of the meal hours. This was a much less formal approach than was the case on the DE's. On those ships, dinner was much more formal, and the officers were expected to attend the specified hour and await the arrival of the Commanding Officer to start the meal.

While we were near Mururoa, the CTs would have to split their meal times so that there was always at least one [redacted] and one [redacted] to keep things working in the forward chartroom. When away from Mururoa, only one person had to be on duty in the forward chartroom. This usually fell to one of the communicators as they were more proficient in the care and feeding of the crypto gear and teleprinters.

During mid watches there were usually two CTs on duty to maintain communications and to assist in the shredding of classified teleprinter paper. This classified material destruction was an onerous duty for several reasons. The pulverizing shredder we used reduced teleprinter paper and carbon paper to a powder with a consistency somewhere between oatmeal and coarse-ground flour. However, you had to feed the shredder very carefully or it would choke. It was also so noisy that you needed hearing protection. The worst problem was that no matter how carefully you worked, the room would end up covered with a layer of dust that had to be vacuumed up lest it offend the division officer or visiting guests. Since the noise and cleanup were such a pain in the butt, we usually shredded our classified printouts only about twice a week. The frequency of shredding was a balance between inconvenience and the necessity to maintain the minimum amount of classified trash in case the ship was damaged or boarded. We didn't think either of these circumstances was very likely, but the capture of the Pueblo was still fresh in the memory of the Naval Security Group.

The Wheeling had a full medical team aboard which included a nurse-anesthetist and a surgeon. The nurse, Tim M, was my cabin mate in 1972. In addition to the medical officers, there were about 5 other military officers besides commander B. Some were Navy officers, others were in the Air Force. One of these, major Dix P, was my cabin mate in 1973. I've forgotten the names of the others, as I seldom saw them outside the wardroom. Commander B normally ate with the Captain and the senior SRI scientists in the Captain's mess.

While we were setting up the forward chartroom, technicians from SRI and the other contractors and DOD agencies were installing their equipment and attaching vans and radar antennas on the Wheeling's side decks and aft helicopter deck. The ship's crew also welded several refrigerated trailers to the side decks to hold frozen food. The Wheeling's freezers weren't large enough to hold frozen foods for a three-month cruise with a full complement of technicians. The ship's chief steward apparently had a generous budget, as the Wheeling never ran out of steaks, prime rib, ham, or chicken.

Before we left Port Hueneme, all the civilian and military crew got a briefing on the mission. We also had to report to the medical center where everyone got an EKG. Apparently, they were afraid of heart attacks during the long deployment far from friendly hospitals. None of the NSG detachment had problems, but perhaps a few of the older and heavier civilian crew had to be replaced.

In early May of 1972, the Wheeling left Port Hueneme for Hawaii where there would be some coordinating meetings with the Air Force and where we could pick up any last-minute additions to the crew or equipment. The five-day trip to Honolulu also had a first for the Wheeling: several women were aboard. The women were technicians and scientist from SRI and were doing some last-minute adjustments to their equipment. They would not be going along on the mission to Mururoa, but they did get a taste of what life at sea would be like for the people on the Wheeling.

When we arrived in Honolulu, we picked up a few replacement pieces of gear and key cards for the crypto gear. I called a few friends to let them know I would be gone for a few months. I left my car keys with a friend at NAVCOMMSTA, along with a letter saying he had my permission to use the car as he wished. I think the car was used by a newly-reporting officer for a week or two until his own car arrived from the mainland. I also left checks with my landlord to cover rent and utilities for a few months.

The Wheeling left Honolulu for Mururoa in the second week of May. It took us about seven days to steam the 2800 miles from Oahu to Mururoa. During the transit the CTs made sure all the gear was working.

When we arrived at Mururoa, we approached to about 12 nautical miles from the west end of the atoll. From the upper decks of the Wheeling we could see some of the larger structures on the island. We were approached by a French minesweeper curious about the large vessel with many radar dishes. The captain notified them that we intended to stay outside the 12-mile territorial limits but would remain in the area for some time. I later found out that the US military attaché in Paris had notified the French government of our intent to observe their tests. France had declared a 200-mile safety closure zone around Mururoa and a larger area downwind where fallout might be expected.

The Wheeling had an excellent inertial navigation system which it could update from satellites or long-range navigational aids. However, the system would become unstable if the ship stopped and just bounced around in the seas. As a result, we adopted an elongated race-track course to the west of Mururoa. After the first few turns on this course, I found that our VHF ████ antenna, mounted on the fore deck, would lose signals if we turned away from the island and the ship's superstructure blocked the signals to the antenna. I discussed this with the Captain and commander B and we settled on a figure-eight pattern where the ship always turned toward the island at the end of each leg. This allowed us to maintain uninterrupted coverage of ████████. We also found that the high-powered HF transmitters aboard the Wheeling interfered with our own HF ██████ capability. After some negotiation, we settled on a schedule that allowed the ship to transmit for a few hours near the middle of the day and for most of the night. HF communications usually worked better at night, so this wasn't too much of an issue. We did have a bit of resistance from Captain Gibson, as he was an avid amateur radio operator and often used his equipment to provide phone patches for the crew when conditions permitted. He was usually able to get a link back to California for at least a few hours per day. He had the advantage of a large high-gain antenna at the top of the Wheeling's mast—something few HAMs ashore could afford.

Both the CTs and the ship's radiomen would often tune a receiver and dedicate a teleprinter to AP, UPI, or BBC HF radio links. The copied news printouts would be posted on a bulletin board outside the radio room. The CTs also copied and printed out encrypted fleet intelligence broadcasts. These printouts were usually less interesting than the unclassified news broadcasts, but we copied them to keep up with Navy changes in ship movements and policy. There wasn't much real value to these as we were usually the only Navy vessel within about a 3000-mile radius. We also got regular classified message traffic from NSA, NAVCOMMSTA Honolulu and the other commands involved in the Hula Hoop project.

French Naval encounters

The Wheeling got several visits from French patrol craft each summer. Minesweepers were the patrol craft we encountered most often. The first few encounters resulted in some maneuvers at less than 100 yards as the French patrol craft would try to take advantage of the rules of the road to try to make the Wheeling change course. The usual result was that Captain Gibson would slow the Wheeling to about three knots and hoist a signal indicating that the ship had restricted maneuverability. He would then continue on his course and the much smaller French craft would sheer off—with much waving and picture-taking on both vessels.

Early in 1972, we received a visit from the Comte de Grasse, a French cruiser that was the mobile command ship for test center. The captain of this ship didn't try to play any games with the rules of the road. He simply steamed by about a quarter mile away, then speeded up and went on his way. I suppose the Admiral in charge and some of the senior officials on board the de Grasse simply wanted a closer look at the large white ship with all the radar antennas.

Test Days

The sequence of events preceding a nuclear test was fairly consistent for tests at Mururoa. A day or two before the test, the French technicians would inflate the balloon that would suspend the device above the atoll. The balloon would raise the device from 500 to 900 feet above the lagoon and atoll. This was done to keep the fireball from touching the surface. If the fireball stayed above the surface, the debris cloud would include much less irradiated water and surface material. I suppose that it would have been possible to estimate the expected size of the explosion by measuring the height of the balloon above the surface, but I never heard that anyone made such predictions. As soon as the balloon was spotted, word would get around the Wheeling: "The Balloon is up!" At some point, nearly everyone found a pair of binoculars or a telescope to check out the balloon.

The balloons used to raise the devices were much like the barrage balloons you've seen in the movies or the Goodyear blimp. They were about 80 feet long and 40 feet in diameter. From 12 miles away, they didn't look very large but they were clearly visible with the naked eye. With a good telescope (and there were lots of those on the Wheeling), you could even see the sunlight glinting off the mooring cables and the box underneath that contained the nuclear device.

On a few occasions, the balloon would be pulled back down, probably to fix some technical problem. Generally, once the balloon was up, we could expect a test within a day or two. The scientists and technicians aboard the Wheeling would transition from their slow-paced life between tests to more active preparations for the test. In the forward chartroom, we would send a message noting the balloon raising to all our intelligence consumers and set up for round-the-clock [redacted]

It was the French practice to evacuate all the non-critical personnel from the lagoon aboard the naval vessels and supply ships. They would generally start this evacuation the evening before the test. Our ability to confirm this evacuation was one of best intelligence indicators of an impending test. [redacted]

When the [redacted] went out, the ships would start leaving the lagoon and move to a staging area north of Mururoa. We would keep the Wheeling well out of their path but could still observe the evacuation on the ship's radar. The actual ship movement might not be visible on radar for an hour or two I would generally try to get some sleep from about midnight to 6AM. During the night, the Wheeling would move out from the 12NM racetrack course to a position about 20NM from the test site.

The morning of a test, the balloon would be visible, however most people on the Wheeling would avoid looking at it in an almost superstitious fashion. Nobody wanted to be looking at the balloon should the bomb go off unexpectedly. In the forward chartroom, we would be monitoring We would relay the expected time of detonation to Commander B and he would make announcements over the Wheeling's public-address system.

When the countdown reached about 15 minutes, our technicians would start their high-speed cameras and recording gear. It then became very important to immediately pass on any indicated delays in the test. The cameras and recorders had only about a half an hour of film and tape capacity, and a delay could mean that they would be caught in the middle of changing film or tape when the bomb went off. There were only a few tests where there were delay in the last minutes of the countdown.

Sometimes there might be a burst of static on our radios—or we just imagined it and it was the exhaled sigh of relief of the French communications officer doing the countdown.

As the detonation occurred, there would be a confirmation on the PA system. At that time one or two of the CTs or myself, would be allowed to rush up to the main deck to observe the mushroom cloud and wait for the "BOOM". We were 20NM from the blast and it would take the sound 100 seconds to reach us. That gave us time to get up to the deck from the forward chartroom. For some of the smaller tests, the result could be a bit anticlimactic—a bit like distant thunder. For other, larger, detonations, there would be a visible mushroom cloud and a very loud "BOOM".

For some of the tests, a few CTs were able to don darkened goggles and stay on deck for the detonation. They would face away from the detonation until the countdown reached zero and they saw the flash reflected of the white bulkheads of the Wheeling. They could then turn around and perhaps see the actual fireball. Since I was the designated [redacted] I didn't get to do this. I usually went up to one of the optical vans later in the day when the scientists would hold a public viewing of one of their video recordings. These were more interesting than live viewing, since the detonations were recorded over multiple optical bands from the infrared to the ultraviolet. The cameras were on stabilized mounts with long lenses that got really good slow-motion videos. Since we didn't see any really large tests (nothing above about 13 Kilotons), they all looked pretty much the same. In 1971 and 1974 there were tests over 10 times larger at 160Kilotons. I would like to have seen those videos.

This engineer looked like he had come straight off the hippie streets of San Francisco. He had a bushy, dark hairdo and full beard. He wore shorts, flip-flops, and a tie-dyed T-shirt. I soon learned not to make assumptions about technical skills from dress or grooming choices.

I watched his prototype construction with fascination. He did not lay out his parts on a prototype board in the usual fashion that I used with my own electronics projects. Instead, he started with a cable carrying the extracted video signal hung from the ceiling with alligator clip test leads. Other wires carrying power came in from the ceiling also. To these wires he soldered resistors, capacitors, transistors and other components. The resulting circuit started to look like an abstract electronic sculpture. He would solder components in place and look at the signals with his oscilloscope. He would mutter and frown, and tack on another component. This went on for almost two days. The result was about 24 inches from top to bottom and about 15 inches in diameter. At the bottom, a coaxial lead disappeared back into the television.

A picture appeared on the television. It was an exterior view of the atoll showing part of a building and the beach. From the shadows of the palm trees, we decided that the camera was at the east end of the atoll pointed west. The engineer turned a dangling rotary switch and a different picture appeared. This one was some sort of control board with many displays, indicator lights and a large digital clock display.

I turned to the senior SRI scientist and said, “Very impressive, but what am I supposed to do with that?” I gestured to the abstract sculpture.

“That’s just step one. Watch and learn.”

I watched. A second engineer stepped up to the prototype and started taking notes and drawing a schematic.

The SRI scientist said, “Part two should be ready sometime tomorrow afternoon.”

I returned to the SRI van late the next afternoon. The electronic abstract was gone. In its place, there was a prototype board with carefully placed and connected components. A rotary switch mounted to the board selected one of the four views: two exterior views and two different control panels. One of the control panels never seemed to change. I told the SRI leader that I was impressed but the process didn’t fit my expectation of the working procedures for a top-flight defense contractor. He replied that when you’re going to spend a few months far from the lab it pays to bring people who can improvise and to pack lots of spare parts. The two engineers had been working as a team for several years on field projects. They could handle the routine stuff without breaking a sweat. When the unusual requests came along, they really proved their worth. Later in my career, I worked with similar people on oceanographic research cruises. I also did my best to become one of those people.

Summer 2: 1973

During the setup phase of the 1973 Hula Hoop operations at Port Hueneme I observed one of the most interesting operations of my career. We had finished setting up the intercept stations and were monitoring local VHF traffic to test the gear. We copied some VHF voice signals that seemed to be related to early testing of the F-14 fighter jet. (A jet most famous for being Tom Cruise's plane in Top Gun). The F-14 was very new and still in the test stage in 1973. The aircraft was designed to be a fleet defense fighter to intercept and destroy attacking Soviet aircraft. Its primary weapon was the very large and very long-range Phoenix missile. Due to the missile's size, only the F-14 could carry the missile. The Phoenix also required the new and sophisticated long-range radar on the F-14. This was a very complex weapons system and required a lot of testing. The F-14 could also carry 6 intermediate-range AIM-54 anti-aircraft missiles.

The day after we first figured out that the F-14 was being tested at NAS Point Mugu, I got into my rental car and drove over to the airfield to see if I could get a look at the hot new aircraft. I drove around behind the hangars and walked around looking for the plane. I found one parked out on the ramp surrounded by pylons with warning tape and a marine guard posted in front.

Being a young and curious officer, I walked up to the aircraft and started to walk around it to get a good look. The Marine asked, "Sir, do you have clearance to approach the aircraft?"

I said, "No, just curious."

The Marine wasn't sure what to do. He looked like he was about to reach for his radio, so I said, "You know, it IS parked out here in plain sight. Am I your first visitor?"

He said "Yes, sir."

"Well, when your sergeant asks, you can tell him I stayed well clear of the warning tape."

After exchanging salutes, I walked back to my rental car and returned to the Wheeling. I was glad I hadn't brought my camera or pushed the Marine guard past his reporting threshold.

The next day, as we were doing some more testing and equipment setup, we again listened in on the flight testing. This time there was quite a bit more excitement. During an AIM-54 test missile launch, there was a malfunction and the missile pitched up and struck the launching aircraft. The aircraft went out of control and the test pilot and his flight officer had to bail out. We listened with rapt attention as the range communications became intense and excited and rescue helicopters were launched. The aircraft was lost but the pilot and flight officer survived. From that day forward, the pilot had to live with being the only Navy pilot to shoot himself down with his own missile.

Another Ship Added

The 1973 Hula Hoop mission was significantly more complex that the one the preceding year. Additional experiments involving the testing of radar transmission through the nuclear fireball and cloud were planned. In addition to the Wheeling, an aging helicopter repair ship, the USNS Corpus Christi Bay, was assigned to the operation. After many years' service in Vietnam, it was in such poor condition that it had to pull into Papeete for emergency repairs before sailing south to Mururoa. In the harbor there, several tanks and void spaces were filled with concrete to stop serious leaks. The CC Bay carried four or five Navy helicopters, which would drop calibrated radar targets on the opposite side of the nuclear explosion from the Wheeling. The radar targets were aluminum spheres about 18 inches in diameter. Spheres were used because, no matter how they spun, they always had the same radar cross section and would present the same radar target as they fell. The radars on the Wheeling could track the spheres as they were occluded by the nuclear explosion and cloud allowing the scientists to determine the extent to which the radar tracking capability was degraded by the explosion.

This new mission would add a lot of extra responsibility for Commander B. Now he not only had to manage the military aboard the Wheeling, but he had to act as air traffic controller for the Navy helicopters.

New Reporting Requirements

Just as the Wheeling arrived at Mururoa in June, we were given a new reporting task. We were asked to forward daily situation reports to CINCPACFLT, the Director of the Defense Nuclear Agency, and Henry Kissinger at the White House. This was the first time we had been asked to forward our reports directly to the White House. We were curious, but orders were orders and I started what I called the SIGINT Daily Sitrep. At first, we struggled to find material to report and there were days where the report was just "Nothing new to report." Within a few weeks we developed some new sources and our standard monitoring started to yield some interesting information about new activities—particularly the approach of protest vessels arriving from the US and New Zealand.

We were curious why the White House wanted first hand reports from Mururoa. That curiosity wasn't satisfied until just a few years ago when declassified documents indicated that Dr. Kissinger and the White House staff were engaged in covert efforts to speed up the French effort to develop a submarine-launched nuclear missile. They were looking for another nuclear ally to act as a deterrent to Soviet aggression. The Limited Test Ban Treaty and other nuclear non-proliferation agreements prevented the US from openly aiding French development of nuclear weapons, so the US provided "negative guidance". In other words, the French would tell our nuclear scientists what they were trying, and our scientists would tell them if there was a better way to do it—without telling them exactly what to do.

Killing time

The ship's crew and technicians had a lot of time to kill between tests. Many of the civilian technicians spent a lot of time in deck chairs working on their tans. There were ping-pong tables and an electric slot car track in the forward hold. The ship showed movies in the evenings, but I seldom attended as I was usually preparing my daily reports after supper.

We had ship-wide parties on the Fourth of July and when we crossed the equator. Although the Wheeling was technically a dry ship, lots of tape-head cleaner (95% ethanol) was consumed on these occasions. I got the worst acid indigestion of my life on the Fourth of July in 1973. I mixed a pretty strong drink of cranberry juice and 95% ethanol. About 5 minutes after I finished the drink, I felt like it was burning a hole in my stomach! I had to track down Nurse Tim and get him to open up the sick bay, so I could get a dose of antacid. To this day, I still don't like cranberry juice. I'm pretty sure a lot of extra aspirin and Pepto Bismol was dispensed the morning after these parties.

A group of the SRI scientists had a regular sunset cocktail hour just after supper. I joined them a few times but stuck with soft drinks. We always watched for the green flash at sunset but never saw it. There were usually too many clouds on the horizon.

Greenpeace 3 and other protest vessels

During the Hula Hoop operations, a series of events occurred that resulted in my sending the most highly classified messages of my career in the Naval Security Group. Several civilian protest vessels had sailed to the Mururoa area to register their unhappiness with the French nuclear tests. There were vessels from both New Zealand and the Canada. The government of New Zealand even went so far as to send two of their naval frigates to the area. There were legitimate concerns about fallout and contamination of the seas in the area.

One of the best known of the protest vessels was the Canadian yacht Vega, later renamed GreenPeace III. This was a 65-foot sailing vessel manned by a mixed nationality crew. Their goal was to sail into the area and get close enough to Mururoa that they would force the French to delay or shut down their testing. We were aware when the GreenPeace III was in the area [redacted] The Wheeling had been ordered to keep out of sight of the protest vessels if possible so that our presence would not become an issue in the controversy surrounding the French nuclear testing.

At that time, the NSG and NSA could copy the communications of US citizens only under extraordinary circumstances, and if there was reason to believe that ordinary coverage included US citizens, the intercepted material had to be given the highest classification: Top Secret Umbra. (I later learned that some very tightly compartmented groups inside the NSA were regularly intercepting the communications of US citizens. The NSA got into a lot of trouble with congress when this project, named Shamrock, came to light,)

On July 1st, 1972, the GreenPeace III suffered a collision with a French minesweeper that was trying to force the yacht away from the test zone. It was towed to Mururoa for repairs and the crew were transferred to Hao, then to Papeete, where they were deported.

On August 15th, 1973, [redacted] we learned the minesweeper had intercepted and boarded the GreenPeace III. The minesweeper noted that there had been some injuries to the crew of the GreenPeace vessel and the boat had damage to its mast and rigging. They were planning to tow the vessel to the Mururoa. I immediately sent a high-priority Top Secret message back to the usual recipients notifying them of this incident. The crew of the yacht were beaten by the French boarders and the captain, David McTaggart nearly lost the vision in one eye as a result of the beating.

The yacht crew managed to smuggle out film showing the boarding and beatings. McTaggart later filed a civil suit that resulted in the release of the photos and evidence that the Canadian government had been complicit in the boarding of the yacht. McTaggart won some small legal victories in 1974 and 1975 that proved very embarrassing to the French government. Press and government sources noted that the McTaggart legal victories were one of the main reasons that France discontinued atmospheric nuclear testing after 1974.

[redacted] So far as we knew at the time, none of these vessels had sighted the Wheeling, although they may have known there was a US ship in the area. I later read in a French book, "Atolls de l'Atome"

(The Atomic Atolls), that the Wheeling had been sighted, but was too far away to identify.

Another protest vessel, the Fri, [redacted]

From the Wikipedia page on the yacht Fri:

On March 23, 1973 Fri sailed from New Zealand into open ocean towards Mururoa. At Mururoa Fri maintained a 53-day vigil within the test exclusion zone, just outside Mururoa Atoll and in sight of the test island, with the company of a second peace yacht from New Zealand, the Spirit of Peace, for five weeks. For many weeks, her only contact was by brief radio messages with the New Zealand Government protest warship HMS Otago. In a symbolic act of protest, New Zealand's Labour government of Norman Kirk sent two of its navy frigates, HMNZS Canterbury and Otago, into the test zone area.[10] On July 17, 1973 French commandos stormed the Fri and arrested the crew and ship, impounding ship and crew firstly at Mururoa and then at Hao Island.
Further information: New Zealand's nuclear-free zone

The publicity surrounding the Fri expedition in 1973 and the protest voyages of David McTaggart on the yacht Vega in 1972 and 1973, (McTaggart was severely beaten by French commandos in 1973), made international news and heralded an invigorated protest movement from New Zealand and Australia which eventually forced the French to cease nuclear testing in the Pacific in 1996. The French Military conducted more than 200 nuclear tests at Mururoa and Fangataufa atolls over a thirty-year period, 40 of them atmospheric.

A frigate from the New Zealand Navy was also in the area. It sighted the Wheeling and one of the helicopters from the Corpus Christi Bay. An excerpt from the New Zealand Navy history web page is included in Appendix V.

Radioactive shoes

One of my favorite morning activities was to climb up the main mast of the Wheeling to a small platform above all the radar antennas. SRI had mounted a very good Questar telescope there and I could use that telescope to watch activities at the western Mururoa test sites.

During one test, the Wheeling had accidentally gotten a bit of fallout exposure due to a sudden change in the winds. Our radiologic alarms went off and the ship was sealed up and our wash-down system activated. This system had been installed for just this type of incident. High-volume seawater sprays covered the ship from stem to stern and were supposed to remove any radioactive particles. They also cleaned all the dust and accumulated salt from the decks and bulkheads. After an hour of wash-down, the radiologic incident teams went out with their detectors and declared the outside of the ship safe for normal traffic. The amount of radiation received was minimal and nobody got any significant exposure. Everyone on board wore a film badge on their belt to measure accumulated radiation exposure, and I was never notified that I had received any abnormal exposure.

The one thing I had not realized was that the wash-down system didn't reach up to the level of the observation point where I would watch Mururoa through the Questar telescope. A few days later, I went up for a morning viewing session. When I climbed down the mast, I was met at the bottom by one of the radiologic safety officers (RSOs), Geiger counter in hand. He had realized that I was above the wash-down perimeter and might have picked up some contamination.

Alas, when the detector scanned the bottoms of my tennis shoes and the seat of my pants, it got a bit noisy and I got a bit nervous. I had picked up a detectable, but not serious amount of radioactive contamination. The RSO jokingly asked, "you weren't planning on having children were you"? I must have had a shocked expression and he quickly added, "Just kidding—probably no worse than an X-ray or two.

I really had minimal exposure, however my tennis shoes and uniform trousers had picked up some contamination. The RSO decided that the shoes had to go overboard, but the trousers would be OK after a few special washings.

Tapping the phones

Early in the summer, we discovered a very strong VHF signal which was transmitting continuously. We analyzed it with our spectrum analyzer and found it was broken down into about 16 continuously active subchannels.

One problem is that each of the subcarriers was FM modulated and we did not have the equipment to separate the subcarriers and demodulate them. I used a patch cable to connect the 10.24MHz intermediate frequency (IF) output of the VHF receiver to an R-390 HF receiver. The IF output was normally fed to a spectrum analyzer to display the subcarriers of a broadband signal, or to display a range of signals so that you could view transient signals within a few megahertz of the tuned frequency of the receiver. With the IF signal fed to the R-390, I could tune the HF receiver to scan through the subcarriers of the VHF signal. As I scanned through the subcarriers, I could tell that some of them were being modulated, but the signals were not intelligible. The problem is that the subcarriers were FM modulated, while the R-390 had only an AM detector. The next step was to implement what is called "slope Detection". This technique takes advantage of the slope of the internal filter on the HF receiver to do a crude form of FM detection. When the FM signal frequency is closer to the center of the HF filter, the output is stronger. When it is further from the center of the filter, the signal is weaker. It took an hour or two to find the right combination of filters and HF frequency tuning to get an intelligible voice from the receiver. Once I had found the settings they were written down and I showed how to set up the combination of receivers. We had used a splitter to have the VHF output connected to both the R-390 and the spectrum analyzer and we found that we could tell circuit was active because the spike on the spectrum analyzer become a bit wider and fuzzier. Within a few days we had copied enough signals to find that certain channels were used more often than others and we concentrated on those channels.

Programming lessons for the CTs

During the slow times between tests and when the Wheeling was sailing to and from Pago Pago for resupply, the SRI technicians offered a programming class using the PDP-11 minicomputer in the ship's telemetry control room. The programming lessons occurred in the morning, when I and the linguists were pretty busy. As a result, only our maintenance technician and one of the O-Branchers were able to take the class. I had taken a class in programming in a BASIC-like language during my one semester at OSU, I was jealous of those able to take the class. When I got back to OSU in 1974, one of the first things I did was to find out what computing resources were available to students. By 1975, I was building and programming my own microcomputer systems.

Super Spook to the bridge

By the middle of the 1973 Hula Hoop operation, my position and that of the CTs was pretty well known, if not openly discussed. It was obvious that we were performing some kind of intelligence collection operation. After all, my voice was on the PA system giving a countdown to the tests. There were some embarrassing moments occasionally, particularly when Captain Gibson had a question about intentions. Once, when a French minesweeper was approaching, I heard over the PA System, "Super Spook to the Bridge. Super Spook to the bridge!"

The CTO on duty turned to me with a smile and said, "I think he means you, sir." I quickly hurried half the length of the ship and up about 5 flights of stairs and arrived on the bridge somewhat out of breath.

"What do you think this guy is up to?" said Captain Gibson, pointing to the minesweeper approaching our Port quarter.

I was a bit taken aback. . There was nothing to allow us to judge his intentions.

I replied, "We've got nothing to indicate he's here for more than a look-see. They've pretty much given up on the Rules of the Road games." I suggested that we all wave and take pictures. Captain Gibson huffed and growled for a minute, then conceded that there wasn't much else we could do. The minesweeper cruised past and both sides waved and took pictures.

1972 Tests

June 24 < 20KT
June 30 < 20KT
July 27 < 20KT
July 31 Safety Test

1973 Tests

July 21 20KT
July 28 <20KT
August 18 < 20KT
August 24 < 20KT
August 28 < 20KT (Air Drop)
September 13 Safety Test

USNS WHEELING (T-AGM 8).
c/o Fleet Post Office
San Francisco, California

MENU
SATURDAY
20 MAY 1972

BREAKFAST:

CHILLED FRUIT JUICE ·CHILLED HALF GRAPEFRUIT
CHILLED FRESH RIPE PINEAPPLE SLICES AND PAPAYA HALVES W/LEMON WEDGES
CHILLED PRESERVED FRUIT STEWED PRUNES
HOT HOMINY GRITS CEREAL ASSORTED DRY CEREAL
CORNED BEEF HASH AMERICAN FRIED POTATOES
EGGS COOKED TO ORDER: FRIED BOILED SCRAMBLED POACHED
EGG CUSTARD EN CUP
BUTTERMILK HOT CAKES OR CINNAMON FRENCH TOAST MAPLE SYRUP HONEY
CHEESE OR MUSHROOM OMELET
ASSORTED FRESH BAKED BREAKFAST SWEET COFFEE CAKES
DRY TOAST BUTTER JAM JELLY MARMALADE
COFFEE TEA INSTANT CHOCOLATE POSTUM OVALTINE
FRESH FRUIT IN SEASON CHILLED FRESH MILK

LUNCHEON:

CELERY STALKS SPRING GREEN ONIONS GARDEN RADISHES MIXED SWEET PICKLES
CREAM OF MUSHROOM SOUP BUTTERED CROUTONS STUFFED OLIVES
FRESH GARDEN SALAD GARNISHED W/SMALL SHRIMPS CHOICE OF DRESSING
GRILLED T-BONE STEAK WITH OR WITHOUT FRIED EGGS MUSHROOM SAUCE
CRAB LOUIE SALAD PLATE: FRESH KING CRAB, ASPARAGUS SPEARS, SLICED BEETS, TOMATO WEDGES, HARD BOILED EGGS, LOUIE DRESSING
GRILLED PIMIENTO CHEESE SANDWICH ON WHITE OR RYE BREAD POTATO CHIPS
BUTTERED FRESH CORN ON-THE-COB CREAMED CARROTS AND SWEET PEAS
OVEN BAKED POTATOES W/SOUR CREAM FRENCH FRIED POTATOES
STEAMED RICE HOT GARLIC BREAD
APPLE PIE A LA MODE ICE CREAM W/HOT FUDGE TOPPING
~~MIXED FRESH FRUIT COMPOTE WITH WHIPPED CREAM & CHOPPED NUTS~~
FRUIT JELLO W/WHIPPED CREAM ASSORTED COOKIES PAPAYA A LA MODE
CHEESE: AMERICAN PROCESS AND SWISS NATURAL CRACKERS
COFFEE CHILLED FRESH GRAPEADE CHILLED FRESH MILK
FRESH FRUIT IN SEASON CHILLED BUTTERMILK

DINNER:

CELERY STALKS SPRING GREEN ONIONS GARDEN RADISHES SLICED DILL PICKLES
CREAM OF MUSHROOM SOUP BUTTERED CROUTONS GREEN OLIVES
TOSSED TENDER GREEN W/TOMATOES AND SALAMI STRIPS SALAD CHOICE OF DRESSING
ROAST TOM TURKEY SAGE DRESSING GRAVY CHILLED CRANBERRY SAUCE
BELL PEPPER STUFFED W/CORNED BEEF TOMATO SAUCE
BROILED SCALLOPS SAUCE
DICED TURNIPS W/BACON GREEN BEANS W/ONIONS /BISCUITS
CREAM WHIPPED POTATOES FRENCH FRIED POTATOES STEAMED RICE HOT
JELLY ROLL ICE CREAM W/CHOCOLATE SYRUP
BLUEBERRY SHORTCAKE W/WHIPPED CREAM & CHOPPED NUTS
FRUIT JELLO W/WHIPPED CREAM ASSORTED COOKIES PAPAYA A LA MODE
CHEESE: PROVOLONE AND BLUE CRACKERS
COFFEE ICED TEA CHILLED FRESH MILK CHILLED BUTTERMILK
FRESH FRUIT IN SEASON

SUBMITTED:
C. A. [illegible]
Chief Steward

APPROVED:
E. R. GIBSON
Master

NOTE: FOR DIET PURPOSES, COTTAGE CHEESE AND CHILLED PRESERVED FRUIT OF THE DAY ARE AVAILABLE UPON REQUEST.

Figure 17 A daily menu from the Wheeling
I often wondered how long it took to type this up each day.

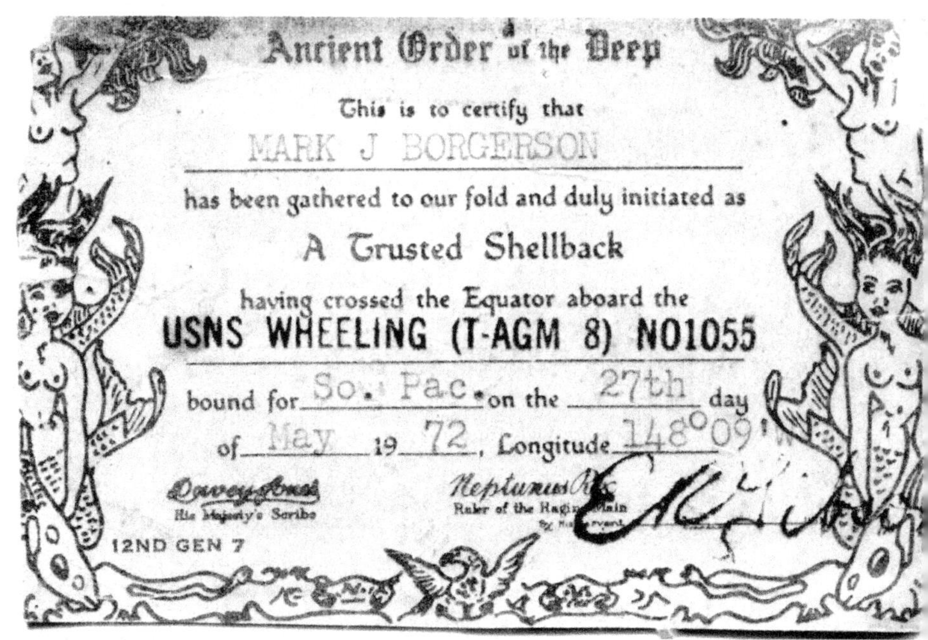

Ancient Order of the Deep

This is to certify that

MARK J BORGERSON

has been gathered to our fold and duly initiated as

A Trusted Shellback

having crossed the Equator aboard the

USNS WHEELING (T-AGM 8) NO1055

bound for So. Pac. on the 27th day

of May 19 72, Longitude 148°09'W

Davey Jones
His Majesty's Scribe

Neptunus Rex
Ruler of the Raging Main

12ND GEN 7

Figure 18 Shellback card from first equator crossing

Figure 19 Posing near the pelorus on the bridge of the Wheeling
Note the lack of mustache. The bag over my shoulder was to carry binoculars and camera up to the top of the mast.

Figure 20 Wheeling at the dock in Pago Pago

The Wheeling was taking on fuel and supplies in the harbor. The photo was taken from a cable car over the harbor.

Figure 21 Helicopter from USNS Corpus Christi Bay delivering mail.

The SECGRU division VHF and UHF antennas are to the left of the fire guard in the aluminized suit.

Figure 22 French Minesweeper Dunkerquoise

This French minesweeper visited Pago Pago at the same time as the Wheeling in 1972. This is a sister ship of the vessel involved in the collisions and capture of the Greenpeace II and Fri.

Figure 23 The Helicopter repair ship USNS Corpus Christi Bay.
The CC Bay served for many years as a floating repair facility for Army helicopters in Vietnam. It was decommissioned in 1975.

Chapter 14 When was our last haircut, sailor?

When the 1973 test series ended in September, everyone on board was in a hurry to return home. Captain Gibson plotted a direct course for Port Hueneme and ordered chief engineer to get the most he could from the engine. While we steamed at about 17 knots most of the trip, it still took about a week to return. While we were en route we packed up as much of our gear as possible. When we finally made port, the CTs were ready to hit the EM clubs, but first they needed to stop by the base disbursing office to collect some pay. I went along as custodian of the payroll records—and to collect some money myself. As we left the payroll office, the CTs were stopped by a lieutenant commander who asked, "Who's your department head? You men are a disgrace to the Navy. You all need haircuts and some new dungarees."

He was correct about their appearance. It was pretty scruffy, when compared to the other sailors around us. The men had been working amongst civilians and haircuts had not been high on the priority list. You can see that in the detachment photograph. Their dungarees had seen three months of wear and, despite frequent trips to the laundry, had stains from carbon paper and the dust from the paper shredder

One of the more brazen 2nd class petty officers replied, "Sir, we've been on a civilian ship on a classified mission for three months. We were supposed to minimize the military presence on the ship. If you have more questions, you can contact Commander B on the Wheeling."

At that point, I approached the LCDR and said, "I'll make sure the next stop is the barber shop, sir." The LCDR let us go with a stern look, and we heard nothing more about the incident. We did all go to the barber shop, realizing that we would need to clean up our act before reporting back to TGU, where some of the chiefs could be pretty strict. Even the sub riders, notorious for coming back with beards and long hair, had to clean up when they came ashore.

Figure 24 SECGRU Division on the USNS Wheeling
You can see why the CTs later got told to get haircuts

Chapter 15 Hula Hoop Post-Mission Briefings

After the end of the end of the 1973 Hula Hoop mission, I was asked to give a classified briefing on the mission for the admiral commanding the Third Fleet and his staff, with special attention to the effectiveness of the direct support mission. I spent about a week typing up my notes, making up a few transparencies for overhead projection, and sorting through the slides I'd taken to select some for visual aids for the briefing.

I talked about success in predicting the tests and delays that might have caused problems for the Air Force. I discussed hardware hack and how it had provided a lot of information. I showed slides of the Wheeling, some French ships, and a few pictures of distant mushroom clouds.

After the briefing, a staff LCDR asked to keep the slides and transparencies to make copies for their records. He said that the staff was impressed by the briefing and I could expect a commendation message. While I later got the commendation message, I never saw my slides again.

In October, TGU and CincPacFlt got a message from the director of the Defense Nuclear Agency requesting my presence at a post-mission roundup meeting in Washington DC. This apparently caused a bit of a stir and some jealousy at the staff level. The only invitee from CincPacFlt was a lowly Lieutenant Junior Grade!

I got travel orders for a round trip flight to Washington, four days hotel accommodations and a rental car. I flew to San Francisco on a Western Airlines champagne flight and changed planes for Washington. There was no champagne on the second leg and I slept most of the way.

I arrived in DC, got my rental car and checked into the hotel the night before the first meeting. When I arrived at the meeting the next morning, I met Commander B, Major P, and the two senior SRI scientists from the Wheeling. I was also introduced to several officials from the State Department, the CIA and the Defense Nuclear Agency. I felt very much like a small fish in a large pond.

At a morning meeting on intelligence sources and requirements, open only to those with SIGINT clearance, I gave an abbreviated version of my Third Fleet briefing, without the visual aids. The Air Force attendees seemed very happy with the level of intelligence support they got from the Wheeling. There were some other meetings on the general results of the Hula Hoop operation. I attended a few, but don't remember anything about them. A lot of the data analysis was still going on, so the information was only preliminary and most of it was over my head. There were also some sessions discussing the characteristics of the French nuclear weapons for which I didn't have the appropriate clearance. At the end of the meetings, Commander B asked, "So how does it feel to be the most junior officer in the room?"

I told him, "I got some practice at Third Fleet. It's no big deal for us Super Spooks!"

(OK, I'm writing quotes from conversations that occurred 45 years ago. You've got to allow me some literary license. Perhaps I should precede these paragraphs with the standard Navy Sea Story Warning: "TINS" for This Is No Shit.)

I think the head of the Defense Nuclear Agency was impressed with the briefings from the Wheeling crew. Although I didn't find out for several months, Commander B, Major P, and I were written up for the Joint Service Commendation Medal shortly after the meeting. I had the distinction of being a junior Naval officer who was awarded a medal by an Air Force general.

Chapter 16 Pony Express

Pony Express was the code name for a DOD program to collect information on Soviet missile tests. There were several components to the program. Two large tracking ships, the USNS Arnold and the USNS Hoyt Vandenberg used sophisticated telemetry antennas and radars to track the missiles and collect telemetry in the mid-course and reentry phases. Land-based radars in the Aleutian Islands also tracked the missiles in the mid-course phase. Aircraft from Japan and Hawaii would also attempt to collect telemetry signals. One or two specially-configured Destroyer Escorts (DEs) from a group of four stationed at Pearl Harbor would attempt to collect re-entry telemetry and retrieve any floating debris in the target area. I was assigned as one of the NSG division officers for these DEs. The DE portion of the Pony Express operations was also called POINTED FOX, as noted in one of my fitness reports. We never used that name and referred to all the operations as PONY EXPRESS.

The collection of data from Soviet missile testing had a very high priority in the early 1970's. The Soviet Union was making rapid improvements in its ballistic missile arsenal. During this era the US launched its first geosynchronous-orbit telemetry intercept satellites. The Navy converted four Claud Jones class DEs to collect telemetry in the mid-Pacific impact zone. A particular concern for the Navy was the improved capability of the Soviet Submarine Launched Ballistic Missiles (SLBMs). In the early 1970s Soviet SSBNs carried the SSN-6 missile which had a maximum range of about 1300 miles. This mean that the SSBN had to sail quite far from its base to be within range of targets in the continental United States. The US Navy had become quite adept at tracking these submarines through HFDF and other means. In 1970 and 1971 the Soviets started testing the SSN-8 missile, which had a range of 4200 miles. This was a tremendous shock to the US Naval establishment. The longer range of the SSN-8 meant that the Soviet SLBMs could strike the United States from "bastions" of Soviet-controlled waters near their Murmansk or Petropavlovsk bases.

Furthermore, in 1973 and 1974, the Soviets began to test the SSN-18 missile which had multiple warheads.

The Claud Jones class DEs, were built in the late 1950s. The ships were round-bottomed, diesel powered, about 300 feet long, and mounted two 3" guns. During 1969 and 1970 the ships had received a major upgrade which included the addition of a special electronics compartment and several million dollars' worth of telemetry intercept equipment. The telemetry intercept compartment was just aft of the ship's radio room and connected to that room with a door having an electronic lock. Only a few of the ships company, among them the captain and executive officer, could enter the telemetry room during operations.

The four 1033-class DEs were:

USS Claud Jones (DE 1033)
USS John R. Perry (DE 1034)
USS Charles Berry (DE 1035)
USS McMorris (DE 1036)

During 1970, they had participated in IVY GREEN telemetry intercept operations for which they later received the Meritorious Unit Commendation. I received the copy of the citation for that commendation via a FOIA request. The citation is shown, as a scan of the original, is shown as Figure 28. The citation doesn't specify anything about IVY GREEN. In 2012 a declassified CINCPAC Command History had this to say about IVY GREEN:

IVY GREEN operations were supplementary collection operations against foreign missile and space activities. Soviet tests monitored by the United States were conducted periodically during 1970 both in the mid-Pacific broad ocean area and in the Kamchatka Peninsula area. Substantially the same PACOM forces participated in these tests in 1970 and had done so in previous years, and once

<REDACTED TEXT>

On occasion ships or aircraft rendezvoused with Soviet observation ships when no actual tests took place.

PACOM forces monitored the deployment of Soviet Missile Range Instrumentation Ships and movement of such ships to probable test impact sites. Usually tests were announced in advance by the U.S.S.R., who asked that certain impact areas be closed to ships or aircraft.

On 5 August CINCPACFLT informed CINCPAC that one of the destroyers observing a test series had reported impacts at ranges of 3,500 yards east and 2,000 yards down range from him.

The original document can be found at:
https://nautilus.org/wp-content/uploads/2012/01/c_seventyvol1.pdf

Since the text has been properly declassified, I have removed the paragraph classifications from the quoted text. Later CINCPAC command histories have no further references to IVY GREEN. Apparently, the program name was changed to PONY EXPRESS sometime in 1971.

I sailed on the Claud Jones and McMorris for Pony Express missions. Two of the DEs were generally in port at Pearl Harbor and on call for Pony Express operations. The other two might be on western Pacific deployments or other cruises. The ships would rotate through the Pony Express operations so that each ship would have sea time for other activities necessary to keep the crews ready for service in general naval operations. Pony Express operations were broken down into several phases which are described in the following paragraphs.

In-Port Time

While the DEs were tied up at Pearl Harbor the ships' crews performed normal maintenance and training. Groups of CTs from TGU would board the ship during the day to train on and maintain the telemetry intercept equipment. The telemetry room was laid out like as shown in the figure at the end of this chapter.

The telemetry intercept positions were arrayed along one side of the compartment. A comfortable chair was bolted to the deck in front of each position. The chairs had to be bolted down because the ship would roll as much as 30 degrees in heavy seas.

Each telemetry position looked much like the example in the NSA Cryptologic Museum exhibit on Telemetry Intercept. The primary radio receivers are at the bottom of the position. The oscilloscopes above the receivers were used to display the signal characteristics. The operator could also recognize various type of telemetry signals by the sound made when the signals were routed from the receiver through an AM or FM detector and played through headphones.

One test given to new operators was to set up the position to intercept and display a standard TV signal. This could be done because the Tektronix oscilloscopes at the position were very versatile instruments. You could set up the vertical scan to match the 60Hz TV vertical sweep. You could then set the horizontal sweep to match the horizontal scan rate of the TV signal. An FM detector in the position was used to generate a signal proportional to the TV luminance signal. That signal was fed into the Z-axis or brightness input of the oscilloscope. When everything was set up properly, you got an intelligible TV picture in shades of CRT green.

It took me several hours of practice before I could pass this test. The most difficult part was the setting of the sweep triggers to achieve a stable picture without rolling or tearing. I division officer to pass the test. After I passed the test and showed some general competence with other pieces of equipment, the Chief presented me with my very own tweaker. (A tweaker is a small screwdriver used to adjust some of the maintenance settings of the equipment). This was a rare honor. A common question at the time was "What is the most destructive thing on any warship?" The answer was "an ensign with a screwdriver."

Alert Phase

The alert phase usually started when I walked into the TGU office and someone would announce "The SMRIS are out". This meant that the Soviet Missile Range Instrumentation Ships had left their home port and were headed to an impact area for a missile test. The notification might come from one or more sources:

- a surveillance aircraft flying from the Aleutian Islands, Korea, or Japan
- SIGINT intercepts or HFDF tracking from US Navy sources
- SIGINT from US Air Force Security Service stations or detachments on tracking ships
- Satellite communications intercepts of increased radio traffic in the launch area

Once the alert was formalized with orders from the Commander, Naval Security Group Pacific, we would start loading our gear aboard the designated DEs. I would try to call someone in the theater group and my girlfriend to let them know I would be out of town for a week or two. I would then return to my apartment and pack my seabag and leave a note and a rent check for my landlord. These short-notice deployments were the main reason I kept no house plants and never considered getting a cat or dog.

While I was getting ready to ship out, the ship's crew would be busy loading last-minute supplies and preparing for sea. CTs with families ashore would be packing and saying their goodbyes. Those without families were doing the same things I did—notifying friends and postponing leisure-time activities.

One unusual aspect of the last-minute supply runs was that the officers aboard the ship were not fed from the general mess supplies, but from supplies purchased with funds from the officers' Basic Allowance for Quarters (BAQ). The stewards working in the wardroom would purchase supplies with these funds and were responsible for the storage and cooking of the provisions. A complicating factor was that many of the ship's officers lived ashore, in family housing or the BOQ, when the ship was in port. While in port, their food allowance went to the BOQ mess or to feed their families. As a result, only a minimum amount of non-perishable foods was kept aboard for the wardroom. As soon as we got mission orders, the ship's supply officer would hit up each of the officers for a contribution to the wardroom mess. Most officers, particularly the married officers, were not inclined to make more than the minimum required contribution, so the initial stores purchased for the wardroom were not generally as plentiful or varied as the provisions in the crew mess. If the mission extended more than a few weeks, wardroom supplies could get a bit short and the stewards would have to purchase emergency rations from the ship's supplies. The ship's supply officer heard complaints about this, both from the captain and from the chief mess cook. A draw like this generated a lot of extra paperwork for the cooks, stewards, and supply officer. If the ship happened to pull into Midway Island for refueling, the chief steward would request an emergency cash draw and head to the Navy Exchange for supplies. However, this could be frustrating, as the supply situation at Midway could be almost as bad as that aboard the ship.

The CTs would bring aboard several footlockers with classified materials: key cards for the crypto gear, TECHINS, magnetic tapes, spare parts and consumable supplies such as teletype paper. Storage space in the telemetry room was limited, and a lot of the gear was stowed behind the equipment racks.

As soon as I came aboard, I would report to the captain to make sure we had both received the same orders. After that, I dropped my sea bag in an unoccupied bunk in the JO Jungle—a berthing space with four bunks, some lockers, a desk, and a head. I would unpack later—when I was sure that we weren't going to be recalled after a day or two. I would then head up to the Telemetry room to make sure that communications were up and running and there were no last-minute issues to resolve.

While I was reporting in and getting organized, the detachment Chief Petty Officer would supervise the assignment of the CTs to berthing areas and emergency stations. The latter part was simple: in any sort of emergency, the CTs would report to the Telemetry room. The goal, from the viewpoint of the ship's crew was to get the riders out of the way so that they could handle the problem without tripping over bewildered CTs. The chief would also make sure that CTs were assigned to cleaning duties in the berthing area. We would usually ask if the radio room adjoining the telemetry room needed any help getting ready for deployment. If they were in a crunch, they might get a CTO to help set up circuits and crypto gear. The Pony Express missions often required more communications circuits and crypto gear than normal operations, so CINCPACFLT might assign a few extra radiomen to assist the ship's company. Working the communications required a Crypto security clearance, so they couldn't assign random extra personnel from the ship's company.

Transit to Operations Area

Once the ships left Pearl Harbor, the SECGRU division CTs unpacked their gear in their assigned berthing areas and reported to the telemetry room for a general mission briefing. We next unpacked our chests of TEXTA, TECHINS and crypto keycards. The O-Branchers worked with the ship's radiomen to get our secure communication up and running. We also unpacked our boxes of snacks and stowed sodas in the overhead air conditioning ducts. The sodas had to be well secured to keep them from rolling around and possibly disappearing into the ductwork. When that was finished, we started practice and training sessions for the telemetry operators while the [illegible] intercept operators manned their positions.

Training now had an extra degree of realism as the CTs had to cope with the ship's motion, noise and vibration.

One of the most difficult parts of the training for the CTs was learning to rapidly acquire, recognize and properly tune the telemetry signals. Speed was of the essence, since final portion of the missile trajectory lasted only a few minutes. There was also an interruption in the telemetry signals during reentry when ionized plasma formed around the warhead and blocked the radio signals for a minute or two. Reacquiring the signals after this blackout was critical as some of the most interesting telemetry was at the end of the trajectory.

We had some training tapes we could play back on the high-speed recorders to familiarize the operators with the telemetry signal sounds and their appearance on the oscilloscopes. However. that didn't help the operator with the critical step of finding and properly tuning the radio receivers. With the help of my CTTC, I came up with a workaround that helped the operators get the full telemetry intercept experience. The test position at the end of the room had an amazing array of signal generators, mixers and amplifiers. We set up one signal generator to mimic several of the Soviet PWM (Pulse Width Modulation) and PPM (Pulse Position Modulation) telemetry signals. When the Chief agreed that they sounded about right through the headphones, we fed the signals into the modulation input of a VHF signal generator. We could patch this signal into the RF inputs of the telemetry intercept positions with varying RF frequencies so that the operator would have to find and properly tune the signal. The problem with this method is that the telemetry signals were strong and clear and the only signals that showed up on the receiver.

I took this method one step further and suggested that we broadcast our simulated telemetry signal through a spare antenna at very low power. We weren't really set up to broadcast signals. However, even with antenna mismatches and a very low power RF signal, I thought there would be enough signal for a test. We didn't want to test this in port, as even low-power signals on known Soviet frequencies might draw some attention from other ships. We certainly didn't want to try it when close to Soviet tracking ships during an operation as it would reveal too much about our knowledge of their operating frequencies and telemetry modes. We finally tested this training method out while at sea in transit to a collection operation.

The simulated signal training worked surprisingly well. The operators got to work through several reentry scenarios in an hour. They had to use their receivers with their usual antennas. The chief and I could vary the signal strength and frequency and could even simulate the plasma blackout by shutting down the signal for a minute. I don't know whether the CTs on the other DEs adopted this method—we only perfected it on my last operation in February of 1974.

One consequence of this testing of the full signal-acquisition chain was that we found that we had a bad preamplifier in the junction box just below the omnidirectional antennas at the top of our intercept mast. While we had a spare preamplifier, the weather was bad, and our maintenance tech was not only seasick, but had a fear of heights. On the theory that I shouldn't ask anyone to do what I wouldn't do myself, I volunteered to climb the mast and replace the preamp. I reminded the chief that I had earned my tweaker and spent a lot of my teenage years building tree forts sixty feet up in in second-growth redwood trees. I had also recently taken a basic rock climbing course while on leave in Yosemite, so I knew how to harness myself and tie off below the preamp box. The maintenance tech had spent three months with me on the Wheeling the previous summer and was confident that I could accomplish the repair. (He'd had no similar issues on the Wheeling. The Wheeling had minimal roll, the weather was good, and our antennas were only about 10 feet above the deck.)

I briefed the CO and asked for a course that would minimize the ship's roll. Once we steadied on that course, I climbed the mast, tied off to allow me to use both hands, and started the repair. The repair consisted of opening the preamp box, disconnecting and removing the bad preamp and installing the replacement, then closing the box. I think it took me about 30 minutes. Soon after I tied off, I found that the course with minimum roll was directly into the wind and I was getting a strong dose of diesel fumes from the after stack. I yelled down about the fumes and asked for a minor course change. That reduced the fumes but increased the roll.

As soon as I mentioned the fumes, all the onlookers moved upwind in case I puked. That didn't happen, but dropped tools or parts were still a hazard. When the repair was complete, I asked the CTs to verify that the antenna and preamp were working properly. That took about five minutes, during which time I hung on and enjoyed the view.

When I got back to the deck, I was a bit unsteady on my feet and had the jitters from the adrenaline rush. We all retreated to the telemetry room and I had a celebratory Pepsi which had been cooling in the AC ductwork. The sugar and caffeine didn't cure the jitters and I think I was babbling a bit.

It turned out that the omnidirectional antenna and repaired preamp were responsible for a lot of our success in recording missile telemetry later in the mission.

Surveillance

Once we arrived in the operational area, defined as somewhere within sight of the SMRIS, we would cruise slowly around the SMRIS while the bridge crew plotted the position of the Soviet ships. We knew from previous tests that when a launch was imminent, the SMRIS would position themselves in an elongated cross with the long axis along the flight path of the missile. When that happened, we would position our ship in the center of the cross so that we would have the best chance of collecting warhead debris.

While we were watching the SMRIS, they were watching us. They had the advantage that some of them had onboard helicopters. The ones we saw were the KA-25 with two counter-rotating rotors. The helicopter would fly around us while both side took photographs. They would generally stay a few hundred yards from us and not make provocative moves. Unlike other barely-armed surveillance ships such as the USS Liberty and USS Pueblo, we had two 3" dual purpose guns with anti-aircraft capability. There were times when the helicopter would close to about fifty yards upwind, then "bounce" down to about 30 feet in altitude. When they did this, the rotor wash would kick up a lot of spray which would wash over the ship. When this happened, the bridge crew would retreat inside and close the hatches. Luckless sailors and CTs on deck would turn their backs and hide their cameras under their jackets.

We had a few white-knuckle incidents during the surveillance phase of Pony Express missions. The scariest incident occurred on the Claud Jones when the insulation around one of the freshwater evaporators caught fire. These evaporators were low-pressure distillation equipment that used waste heat from the diesel main engine exhaust to produce fresh water for the crew. The insulation started smoldering and the smoke was detected by the engine room crew. They called in a fire alert and the bridge hit the emergency alarm followed by the announcement "Fire in the engine room. This is not a drill!" The crew went to fire emergency stations and the CTs mustered in the telemetry room. With our separate AC system, we never smelled smoke, but it could be smelled throughout ship. The fire never grew to a dangerous blaze, but ALL fires on a ship are taken seriously. This one was identified and controlled in about 20 minutes.

Another incident occurred on the McMorris while we were on station. There was a failure in the hydraulic steering gear during a turn. As a result, the ship could only continue that turn until the steering gear was repaired. As a result, the ship sailed around in a circle for about two hours while repairs were under way. Since the ship could not maneuver, it had to hoist a day signal indicating that it was not under control. (The mnemonic we learned at OCS was "Red over Red, the captain is dead". No captain likes to hoist that signal! Another downside to the situation was that it occurred near lunch time and we could not take a course to minimize rolling during the meal. During that lunch and other meals in rough conditions, the officers got very good at eating with one hand while holding our plates in place with the other. The crew in the enlisted mess actually had it easier. Their food was served on standard metal tray with four areas with raised edges that minimized sloshing. The crew mess was also a deck lower where there was less ship's motion to move the food around. It was almost funny watching the officers handle the soup course: one hand tilting the bowl to keep the soup in place while the other maneuvered the soup spoon.

The 1033-class DEs rolled a lot. They were round-bottomed ships which had been given a lot of new topside weight in the telemetry room and antennas. It was not uncommon to have 30-degree rolls in the large swells of the North Pacific. Seasickness was a common problem—particularly among the CTs who didn't spend that much time at sea and were in a closed compartment well above the roll axis of the ship. Sleeping in heavy weather was difficult. I would often sleep against the bulkhead behind the mattress in my bunk with the mattress wedged against the bunk rail to keep me in place. It was not uncommon to find a few CTs strapped into the bolted-down chairs in telemetry room with their heads on their crossed arms trying to get an hour of precious sleep.

I was seasick to the point of nausea only once on my Pony Express missions. That occurred on a morning where the stewards served very sweet and greasy French toast accompanied by guava nectar. I was braced in a chair in the telemetry room just after breakfast when I realized that breakfast and I were going to part ways. I barely made it out the back door to the deck before losing breakfast. After about five minutes of coughing and spitting, I felt much better. French toast and guava nectar have joined with 95% ethanol and cranberry juice on my Do Not Consume list.

Resupply

There were some missions which extended beyond the fuel and provision limits for the DEs. When this was predicted, a second DE would arrive on station and the first would head for Midway for fuel and provisions. It would take a day or two to get to Midway and we would generally be at the pier there for only about half a day. The stop would give the sailors and CTs a chance to get ashore, visit the Navy Exchange and perhaps a bar at the EM or Officer's club. One of our favorite pastimes was to walk out to the golf course and watch the Laysan Albatrosses (or gooney birds) come in for a landing. While these birds were experts at soaring over the waves, they weren't very good at landing and taking off on land. Midway is a nesting area for these birds and we were supposed to keep our distance from nests and birds on land. It was amusing to watch the birds landing—many would tumble head-over-heels on landing. Unfortunately, a small percentage would tumble hard enough to break their necks and there were often a few dead birds on the grass.

During one approach to the pier at Midway, the captain allowed one of the ship's junior officers to handle the ship. While the crew and CTs watched on deck, he made a nice approach, but didn't slow down in time. We dinged one of the offshore dolphins (groups of pilings about 10 yards from the pier) pretty hard. The dolphin barely survived, and the McMorris got a good dent and needed some repainting of the hull. As a visitor, I stayed quiet in the wardroom that night, but the other officers were not gentle with the junior officer of the deck.

On one mission starting in early 1974, our detachment was aboard the USS Claud Jones at the time Navy-wide advancement examinations were scheduled. We had left Pearl Harbor several weeks before and had not anticipated being at sea so long. We did not have either the exams or the study materials the CTs needed to prepare for the test. CincPacFlt had a procedure to cover this, as it was a common occurrence for the CTs deployed on submarines: a special exam would be administered after the CTs returned to Pearl Harbor.

After we had been on station for about two weeks, another of the DEs, the Charles Berry arrived on station and delivered some critical supplies. I also received a message that the NSG officer on the Berry would be delivering our CT exams to be administered in about two days. This caused great consternation amongst the CTs. They had been tired and seasick for two weeks and had no access to books and manuals needed to prepare for the exam. Remember that this was before the era of internet downloads—the required manuals would fill a footlocker and the manuals were classified and required special handling. We normally didn't take these manuals with us as we had limited storage space and no study area other than our telemetry intercept room. There wasn't desk space to spread out manuals and it was often noisy due to the teletypes in the communications area.

After consulting with my Chief, I sent a message back to headquarters requesting delayed exams and explaining why the exams would be unfair to the CTs. Some lieutenant commander on the staff decided that we didn't rate the same consideration as the CTs on submarines and we were ordered to proceed with the exams. I again consulted with the Chief, asking if there was anything else to be done. He said that if we were ashore we'd be screwed—but we were on a ship. The captain of a ship has nearly absolute control of administrative matters aboard his vessel. I went up to the bridge and pled my case to the commanding officer, Lieutenant Commander H. LCDR H sent a message to the NAVSECGRUDET CincPacFlt staff stating that, since there was a procedure in place for giving alternate exams and there was no secure area aboard the ships suitable for administering the exams, they would be postponed until the ship returned to Pearl Harbor. I'm sure there was some grumbling and cursing amongst the NAVSECGRUDET staff at this countermanding of their order. However, overruling a ship's commanding officer would require taking the issue outside NSG channels and it was unlikely that CincPacFlt would side with the NSG rather than their own commanding officer. The envelope with the exams remained sealed and the CTs breathed a sigh of relief.

LCDR H's stock with the CTs rose to new heights. We had all thought him a good captain. Now we knew he would stand up to staff pressure in the interests of the embarked CTs. I have long felt that he was one of the best leaders under whom I served. After that incident, I was less impressed with the leadership of the NAVSECGRUDET staff. Along with the CTs, I felt that they didn't understand the difficulties of our mission on the DEs, and that they gave unfair consideration to the submarines. We knew that working conditions on the subs were difficult and living and working quarters were cramped. However, the subs didn't roll 30 degrees each way on station!

My own standing with the NAVSECGRUDET staff dropped significantly after this incident. Since I had already received orders to be released from active duty, this was not a major concern to me. I felt a lot more loyalty to the CTs with whom I served than I did to the staff officers who complicated our lives. My fitness report from the Claud Jones was very good. The report from the NAVSECGRUDET staff—OK, but not as good. In the end, it counted for little as the mission reports from the USNS Wheeling, McMorris and Claud Jones were all very good. At that time, the best fitness reports in the Navy would probably not have been sufficient to change the decision to release me from active duty.

While we were on station with the SMRIS, we struggled to keep our high-frequency radio communications circuits open. We were often in what the communicators called "the Black Hole of Midway". HF radio signals bounce off the ionosphere, allowing them to have hundreds to thousands of miles of range. However, there are areas along the signal path called "skip zones" where communications are unreliable. The locations of the skip zone are dependent on the frequency and time of day. The communicators could usually get circuits to work most of the time, but even small dropouts would cause the crypto gear to lose synchronization. Resynchronizing the crypto gear required using a new key card and following a procedure that took up to 10 minutes. One of the critical supplies delivered to the McMorris was several packages of crypto key cards.

Launch and Telemetry Intercept

When a Soviet launch was imminent, keeping communications in sync was critical. Unless one of the large US missile tracking ships was in the area, our only way to get warning of the launch was via HF communications. If the Vandenberg or Arnold was close by, they might be able to get the launch warning via either satellite communications or their own HF links and forward it to us via VHF radio.

We had a high-gain steerable antenna in the "golf ball" housing over the Telemetry room, but it was seldom operational. The antenna was supposed to give us high-gain telemetry intercept capability. The rolling motion of the DEs resulted in too much wear and tear on the aiming system servo motors and they usually failed before we arrived on station. Even if they were working, the servos couldn't move the antenna fast enough or far enough to compensate for the ship's motion in anything other than the calmest of seas. Since we had no satellite communications at that time, we had to rely on HF radio for warning signals.

The launch warning came from other national surveillance assets: ships or tracking stations near the Aleutian Islands, or satellites. In 1970 and 1973 the US launched its first geostationary SIGINT satellites, Rhyolite 1 and Rhyolite 2. These satellites were designed to collect telemetry from launches at the main Soviet launch facility at Tyuratam. The satellites were able to collect telemetry for the launch and intermediate phases of the missile tests, but they could not get good results for the re-entry phase of tests which impacted in the North Pacific near 180E. My own suspicion is that the modification of the Claud Jones class DEs was prompted by the lack of telemetry intercept satellite coverage in the mid-Pacific impact area.

HF Launch warning messages went out to the ships in the impact area at FLASH priority. By the time the message arrived aboard the McMorris, we would have about 15 to 20 minutes for the CO to position the ship and for the CTs to be ready to intercept the telemetry. If there was only one DE on station, the CO would generally put the ship right in the middle of the predicted impact area as determined from the positions of the SMRIS. That would give us the best chance to collect warhead debris and the exact position made no difference to the omnidirectional telemetry intercept antennas.

When we estimated that the missile was about 10 minutes away we would start our high-speed tape recorders and begin searching for the telemetry signals. We had to be wary of starting the recorders too early as a tape only lasted about 15 minutes. We could not risk starting too early and having to change tapes in the middle of a reentry. We had two recorders, but we always wanted to keep one as a backup in case of an equipment failure. We would generally collect telemetry in two phases separated by a blackout interval. The blackout occurred as the warhead re-entered the atmosphere. The very high speed of the reentry vehicle would heat the air around it to become a superheated plasma that blocked outgoing telemetry signals. When the blackout occurred, the Chief would announce "Blackout" and note the time on the audio log channel. The CTs with valid telemetry signals would move their hands away from the receivers lest they jar a knob and shift their receiver frequency.

After a minute or two, one or more of the CTs would announce "Reacquisition" and the Chief would note that on the log and everyone would breathe a sigh of relief. We would continue to collect terminal phase telemetry for several more minutes until the warhead hit the ocean. There would sometimes be an announcement from the bridge of a visible impact. Some warheads also had a sounding and destruction charge that exploded on impact. This charge allowed the SMRIS to pinpoint the impact area with sonar gear. It also destroyed most of the warhead electronics.

After the impact the DE would head to the impact area to search for debris. After the CT's had shut down the recorders and logged the frequencies and telemetry types, they were free to go on deck to observe the debris retrieval. The ship's crew had long-handled dip nets to fish out any debris that they could find. All I ever saw recovered were some charred chunks of Styrofoam. Perhaps the size and curvature of the foam told CIA and Air Force analysts something about the warhead—to those of us aboard the ship it was rather anticlimactic. I think the CO and crew did find it interesting to match wits with the helicopter from the SMRIS to see who could get the most Styrofoam!

Return

The end of a Pony Express mission generally came when the SMRIS headed back to port. The DEs would then head back to Pearl Harbor. On the way back, we would prepare summaries of our collection, make backups of the tapes and clean up the telemetry room. Once the chores were done, the CTs and I would catch up on sleep to the degree allowed by the ship's motion and the need to maintain a communications watch.

On the return trip in November of 1973, we discovered that we were missing a used key card from one of our KW38 crypto units. These key cards were carefully inventoried, and loss of a key card was considered a major security breach. With the capture of the USS Pueblo in 1968, the Soviets gained access to US KW38 crypto machines and could potentially decode any intercepted communications for which they could obtain the key card.

We practically disassembled the telemetry room looking for that key card. The search was to no avail and we returned to Pearl Harbor without the card. We finally decided that the key card must have been inadvertently included with classified teletype printouts which had been dumped overboard. A long Pony Express mission would consume dozens of rolls of teletype paper. When ashore, the burn bags would go to an incinerator. At sea on the USNS Wheeling, we shredded the paper and kept it in bags in a large unused compartment. The DEs had neither a shredder or a secure storage area, so the classified printouts were stuffed into weighted bags and dumped overboard in deep water out of sight of the SMRIS.

On the way back to Pearl Harbor I wrote a report detailing the loss and search for the card. The NSG staff must have been convinced that there was no potential compromise, as no disciplinary actions were forthcoming. I later found out that the Walker spy ring (a searchable term) had been copying Navy key data and turning it over to the Soviets for almost 18 years.

One of the most satisfying moments of my naval career occurred after we returned from a Pony Express mission on the Claud Jones in March of 1974. We had collected almost five minutes of good telemetry signals. The Vandenberg was also in the area, but had some trouble acquiring the warhead with its high-gain antennas in the rough seas. As a result, our data was the best collected for that mission, and we generously sent a copy of our tape to the Vandenberg by small boat. They had specialists in telemetry analysis who could do a quick-look analysis of the data. The Claud Jones then headed back to Pearl Harbor with our original tape.

A day or so after we returned, I got a call requesting my presence at NAVSECGRUDET headquarters for a post-mission briefing. This had never happened before. I checked with Chief and he said that such meetings usually happened when something had gone wrong. Congratulations usually arrived in the form of a Bravo Zulu message. Dissatisfaction was expressed in person.

I showed up at headquarters and was shown to the desk of a staff lieutenant commander. He asked, "How do you explain your team's performance on the last mission?"

I replied, "Well, I think we did the best we could with our omnidirectional antennas."

"So you couldn't collect any data because your high-gain antenna wasn't working?"

"Well we might have gotten more if it had worked" I said.

By now the LCDR was in full snit mode. "The only telemetry I've seen from this mission is this tape from the Vandenberg. I don't like getting my data from the Air Force!" He then waved a large routing envelope containing a magnetic tape. It was the same envelope in which we'd transferred our tape to the Vandenberg, which had just arrived in Pearl Harbor. The people on the Vandenberg had made a copy of the tape and returned ours in the same envelope---just blacking out our To and From blocks and adding their own.

I politely said "Sir, if you look closely, you'll see that the tape you're holding originated on the Claud Jones. We sent a copy to the Vandenberg for analysis. The routing message in the envelope should make that clear."

After a few seconds, during which the LCDR had nothing to say, I said, "Is that all sir?" I departed quietly

TGU CincPacFlt and the Claud Jones later received congratulatory messages from CINCPAC and the NSA for our significant contribution to the Pony Express operation. NSG also got a thank-you note from the Air Force for allowing them early access to the telemetry data from the Claud Jones. I would love to have been looking over the LCDR's shoulder when he read those messages.

About a week later the Chief told me I should steer clear of headquarters for a while. The CPO network reported that a certain LCDR had been forced to eat crow over his premature criticism of the Claud Jones detachment. It's never good to put a superior officer in an embarrassing position. I was just weeks from my release from active duty at that point. I spent most of that time working the CTs at TGU to document the training simulations we had developed.

The View from the Bridge

The Claud Jones, like other naval ships, submits a command history each year which describes the ship's operations for the year. I received the 1974 command history for the Claud Jones via a FOIA request. Here is what it had to say about the Pony Express operations in late 1973 and early 1974:

1974 was a year which saw CLAUD JONES perform several missions which resulted in considerable time at sea. On 31 December 1973, CLAUD JONES received orders to proceed to the Midway Islands for Pony Express Operations involving surveillance of Soviet Units in the Broad Ocean Area. Within 24 hours of receipt of these orders the ship was enroute Midway and plans to participate in Project Teal Gull in Mid-January were modified.

CLAUD JONES remained at sea for 71 consecutive days despite adverse weather and limited support making use of underway replenishments with the USS PIEDMONT (AD17), USS KIWISHI (AO-146) and USS SACRAMENTO (AOE-1), plus two brief refueling stops in Midway. The efforts of CLAUD JONES contributed significantly in making this Pony Express Operation a most successful operation. In addition, Project Teal Gull was modified and conducted while on station, further demonstrating the versatility of CLAUD JONES.

Figure 25 The USS McMorris in 1972
The telemetry room is under the after mast which has telemetry intercept antennas. The rectangular box under the antennas contains preamplifiers. (US Navy Photo)

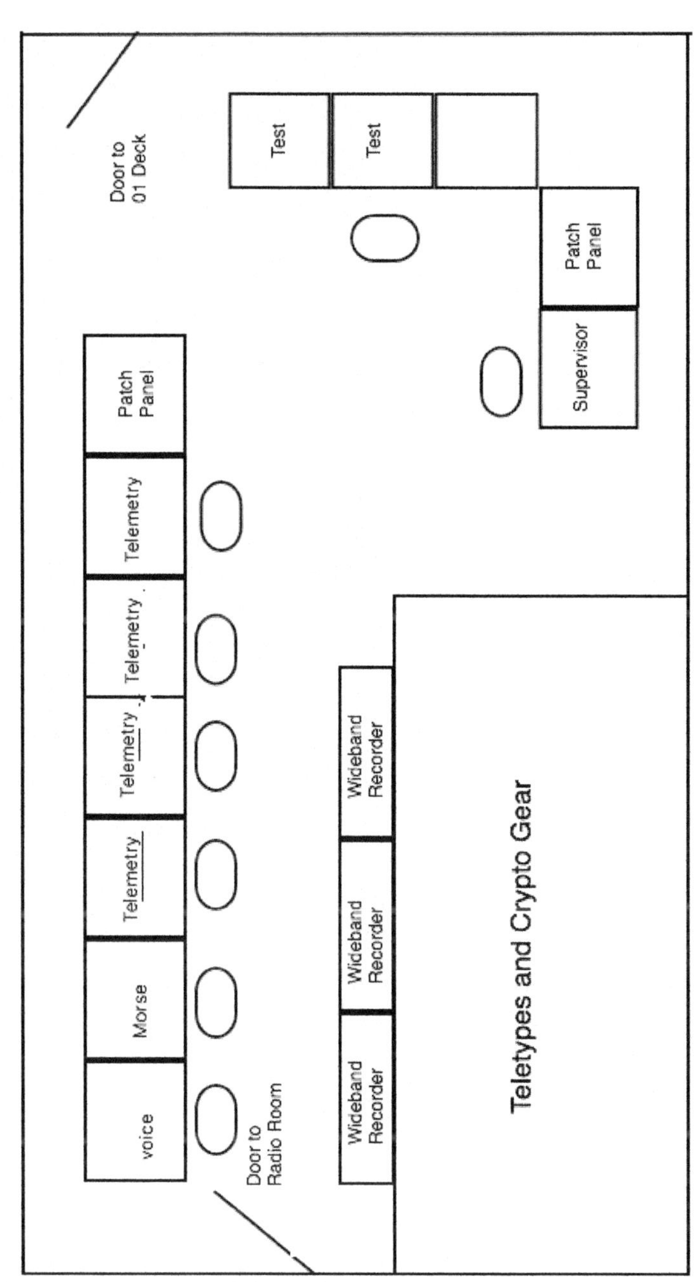

Figure 26 Layout of the Telemetry room.

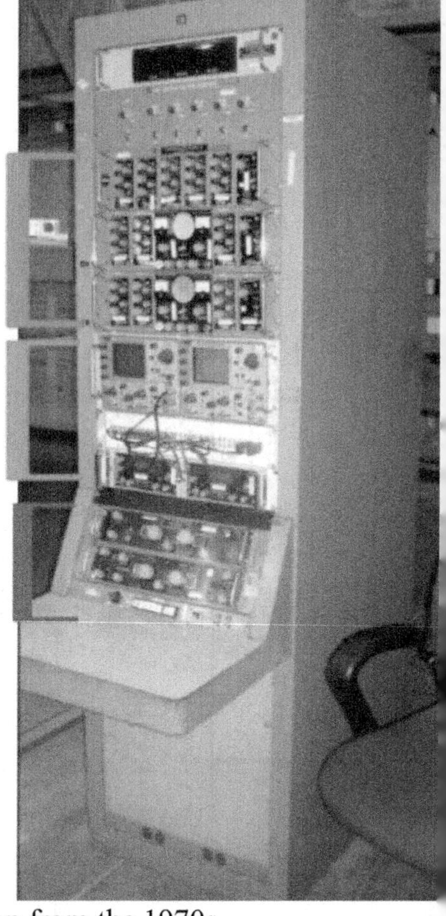

Demodulators: Changed the signals from the receivers into a form that was suitable for recording on magnetic tape. The tapes would then be played back on RISSMAN in the NTPC
Displays: Operators used oscilloscopes to graphically display the signals. This allowed operators to make sure the equipment was adjusted properly, ensuring the signal could be processed correctly.
Receivers: Picked up telemetry from satellites and missiles. These signals contained data on the satellite's purpose or missile's performance, such as reconnaissance photographs or missile trajectory.

Figure 27 Telemetry intercept position from the 1970s
This exhibit in the National Cryptology Museum is nearly identical to those on the Pony Express DEs.

CHIEF OF NAVAL OPERATIONS

The Secretary of the Navy takes pleasure in presenting the MERITORIOUS UNIT COMMENDATION to

TASK FORCE NINETY-TWO

consisting of

Commander Task Force 92 and his Staff; Commander Task Group 92.1 and his Staff; USS JOHN R. PERRY; USS CHARLES BERRY; USS CLAUD JONES; USS MC MORRIS; and Detachments from Patrol Squadrons SIX, SEVENTEEN, TWENTY-TWO and FOUR, and Fleet Air Reconnaissance Squadron ONE

for service as set forth in the following

CITATION:

For meritorious service during the spring and summer of 1970 while conducting IVY GREEN operations in the Pacific Ocean area. Demonstrating alert responsiveness, eager aggressiveness, remarkable flexibility, and unshakable tenacity, the units of Task Force NINETY-TWO carried out IVY GREEN operations which resulted in significant contributions in this most vital aspect of national security. Throughout these operations, all members of assigned Task Force NINETY-TWO units displayed outstanding professionalism, despite many extra and trying hours, with the result that all mission objectives were achieved. The superior teamwork, courage, and skill displayed by the officers and men of Task Force NINETY-TWO attested to their sustained technical competence and unflagging devotion to duty and were in keeping with the highest traditions of the United States Naval Service.

For the Secretary,

E. R. Zumwalt, Jr.
Admiral, United States Navy
Chief of Naval Operations

Figure 28 MUC Citation for Operation IVY GREEN

Figure 29 Soviet KA-25 helicopter near the USS Claud Jones

Figure 30 USS McMorris approaching the Claud Jones

Figure 31 SMRIS Chazma
This is one of the Soviet tracking ships that were always present in the impact area

Chapter 17 Spoofing the RORSAT

In early January of 1974, the Claud Jones embarked on a short mission to evaluate the performance of a Soviet Radar Ocean Reconnaissance Satellite (RORSAT). On December 27, 1973, the Soviets launched the Kosmos 626 RORSAT satellite. This system was a large satellite with an onboard nuclear reactor to provide power to a side-looking radar capable of detecting large ships such as aircraft carriers. Kosmos 626 was the first of the series to have a normal operational profile. The Soviets continued to launch and test these systems until the 1990s. Should there have been a sea war between the Soviets and the US, they would probably have launched several of these satellites to provide wide-area coverage and targeting information for strikes against US maritime assets. Several classes of Soviet ships were equipped with satellite communications antennas to directly receive the data from the RORSATS, which they could then use to guide long-range anti-ship missiles.

Shortly after Kosmos 626 was activated, a representative of the Defense Advanced Research Projects Agency (DARPA) arrived at TGU Pearl Harbor to brief us on a plan to evaluate the performance of the RORSAT. The briefing detailed a plan to use the Claud Jones as a mobile radar target and to collect the telemetry data from the RORSAT. That telemetry data would be collected by NSA intercept of the satellite downlink and provided to DARPA for analysis. A key aspect of the experiment was that a radar corner reflector would be installed on the Claud Jones. A corner reflector is a very efficient radar reflector that returns a consistent and strong signal over a wide range of input angles. (A requirement on a ship that rolls as much as the Claud Jones.) This corner reflector had adjustable panels that could alter the radar cross section of the ship to return a signal to the RORSAT that could appear to be anything from a small destroyer to that of an aircraft carrier. The NSG detachment on the Claud Jones would also record the radar signal as it swept across the ship. The timing of this shipboard intercept would allow DARPA to examine the appropriate part of the intercepted telemetry.

The Claud Jones sailed to the northwest of Pearl Harbor for the tests. For about a week, we collected the radar signals several times a day while the DARPA technician altered the configuration of the corner reflector. We had approximate satellite pass times from NSA and DARPA and we would set up our microwave receivers to capture the signal. When the radar targeted us, it resulted in a very strong signal with a distinctive "brrrp" sound which lasted about one to two seconds.

A report in the November/December 1999 issue of the Journal of the British Interplanetary Society had this to say about the RORSATs:

According to recently declassified CIA reports dating from the 1970s, the US-A radar could "detect medium-sized and some small ships—such as cruisers and destroyers—under favorable conditions, and probably can detect large ships—such as aircraft carriers—even under adverse sea conditions" [36]. Submarine detection was neither an original goal of the system, nor apparently was it possible during operations of the US-A constellation.

I suspect that statement is based largely on the experiments aboard the Claud Jones.

Red Cloud Prototype

During one of the late 1973 Pony Express missions, the USS McMorris carried a prototype microcomputer-based electronic intelligence system code named Red Cloud. This system was supposed to automatically collect and characterize incoming electronic signals over a wide frequency range. The prototype occupied part of an equipment rack near the back door to the telemetry room. It was connected to a line printer that was supposed to print out reports. The equipment came with a civilian support engineer who installed the system and was supposed to keep it running. The project goal was to reduce the size power requirements of the ELINT collection system while providing a hard copy readout of the signals collected and analyzed. Unfortunately, it was a system that went to sea before it was ready.

The Red Cloud prototype used an Intel 8008 microprocessor to control its operation. The microprocessor and its memory and support circuits were mounted on a large circuit board in wire-wrap sockets. After installation and testing at the dock in Pearl Harbor, it went to sea and was tested on the transit portion of a Pony Express operation. Both the electronics and the civilian technician became very unreliable after a few hours exposure to the motion and vibration of the ship at sea. The most visible problem was that the microprocessor and support circuits would pop half their pins out of their sockets due to vibration. The first phase of any test was to open the cabinet and firmly push all the chips into their sockets. Despite this attention, the system performed only marginally at first and not at all after a few days. I suspect that popping half the pins out of their sockets on one side caused power supply issues that fried critical chips. The civilian support engineer's unreliability was due to the amount of time he spent huddled in his bunk suffering from seasickness.

Chapter 18 New Orders and Readjustment

In 1972, I was enthusiastic about my work in the NSG and applied for augmentation into the regular Navy. Augmentation is the process whereby an officer changes from Naval Reserve status to regular Navy status. The primary change is that regular Navy officers remain on active duty indefinitely, while reserve officers can be released from active duty at the convenience of the Navy, usually at the end of a defined period of active duty. This request disappeared into the Navy paperwork maze until August of 1973, at which time I was at sea on board the Wheeling in the south Pacific. I was sent orders to report to the aircraft carrier USS Franklin D. Roosevelt in the Mediterranean in October of 1973. The orders included the provision that, if I accepted the assignment I was agreeing to remain on indefinite active duty. When these orders arrived on the Wheeling, I was in a quandary. At that time, I didn't know when the Hula Hoop mission would end, and I would return to Hawaii. Post-mission briefings could occupy a few weeks, and I was still on the list of officers for Pony Express operations. I wasn't sure I could carry out all those obligations, pack up my apartment and move to a ship in the Mediterranean by the date on the orders.

I sent a message back to BUPERS that I would like to have the orders canceled because of the uncertainties surrounding my current duties and the end of the Hula Hoop mission. That message effectively ended my chances for augmentation into the regular Navy.

After I returned to Hawaii, I gave a briefing to the commander of Third Fleet, who had overall command of the Navy part of the Hula Hoop operation. I presented the details of our signal intelligence operations and showed a number of slides I had taken aboard the Wheeling. The admiral's aide asked to keep the slides for a week or so to make copies. He never returned the slides—perhaps because I was hard to reach because of other travel or TAD on Pony Express operations. On October 19th, I flew to Washington DC to participate in a review of the Hula Hoop program at the headquarters of the Defense Nuclear Agency. The meetings lasted about two days and I returned to Hawaii on the 26th of October.

In the first week of November 1973, I received a letter from the Navy Bureau of Personnel (BUPERS), indicating that I was to be released from Active duty no later than March of 1974. The letter cited the large personnel reductions levied on the cryptologic community at the end of the war in Vietnam. I was aware of this drawdown, as several reserve lieutenant commanders at Pearl Harbor and Wahiawa were also being released from active duty. Some of them had been on active duty for more than 12 years and had been planning to stay on active duty until they reached the 20-year retirement eligibility.

I could see that my active duty time in the NSG was ending but had little time to worry about it. I was at sea for almost two months on Pony Express operations in late 1973 and early 1974. Between that sea time and my work with the Schofield Barracks theater group, I had little time to worry about my future—other than to plan to return to graduate study at Oregon State University in the fall of 1974.

While discussing the plight of the other officers who had been released from active duty, I found out about a benefit they were getting. The benefit was called Readjustment Pay. To be eligible for the benefit you had to have a minimum of five years on active duty and must have applied for augmentation. I met those requirements in January when my total active duty time passed five years—as my two years of enlisted service counted toward the five years. The office worker to whom I presented the request for readjustment pay was a bit surprised that he was doing the request for a LTJG, as most reserve officers didn't pass the five-year mark until well after they were promoted to full lieutenant. (I would have been promoted to that rank in about August of 1974.)

Readjustment pay was calculated as two month's basic pay for each year of active duty, with a maximum of one year's pay. I received 10 months' pay at my basic rate of $937 per month: $9370. When I was released at the end of March, I collected that sum, my final month's pay, and payment for unused leave time. I kept a few thousand dollars for interim expenses and sent my mother a check for $11,000 dollars for deposit to my California checking account. My mom saved that check as a souvenir after it was cleared.

I spent my last few weeks at TGU working on training procedures and other paperwork. On my last day, I was presented with an engraved TGU Wreckers beer mug. One of my collateral duties had been to serve as recreation officer and the Wreckers were the TGU softball team.

On my last day, I reported to my home duty station, NAVCOMMSTA HONO at Wahiawa and was given my DD-214 release document. I turned in my security badge and active duty ID card. Although I visited Schofield barracks several times after that to work with the theater group, I didn't need an ID card as the officer sticker on my car was sufficient to get me in the gate. At that time, I also signed a document pledging to guard classified information, even though I no longer had a security clearance. I was also notified that I was not allowed to travel to certain countries, (China, Cuba, USSR, etc.) for a period of one year.

Chapter 19 After SECGRU

The Fate of the 1033-Class DEs

The 1033-class DEs had a short service life of only 15 years. The official reason for their decommissioning and sale to Indonesia was that they were not suitable for extended anti-submarine operations. This was certainly true. Their maximum operating speed was lower than that of many Soviet attack submarines and they had a limited ASW sensor suite. The decommissioning and sale of the McMorris and Claud Jones, and Perry is a bit more mysterious. They had undergone multi-million-dollar refits to equip them with the telemetry intercept room and equipment in 1971. They were used for Pony Express telemetry intercept operations for only three years before being decommissioned.

It is my opinion that the Claud Jones class DEs were found to be of marginal utility in telemetry intercept operations. The excessive rolling in high seas was detrimental to both operator and equipment performance. Their low transit speeds and limited on-station endurance made logistic support more difficult than was the case with larger ships.

Another factor that may have contributed to the decision to decommission the Claud Jones DEs was that other telemetry intercept platforms may have become available. The second Rhyolite telemetry intercept satellite was launched in 1973 and it may have been positioned to get better coverage of the mid-Pacific impact zone. During the mid-1970s, the Naval Security Group was also beginning to activate temporary intercept facilities in vans or shipping containers placed on the after decks of larger destroyers and cruisers. In, addition, as standard ELINT and COMINT collection equipment became more sophisticated, it may have been able to perform many of the functions of the Telemetry room on the DEs.

Pony Express missions continued to occur for at least a decade after the decommissioning of the Claud Jones DEs. There are references to Pony Express in declassified documents as late as the mid-1980s. The pertinent text has been redacted, so only the mission name is publicly available. As a result, I can't be sure what vessels and equipment participated in these later operations.

Summer in the Bay Area

Shortly after my release from Active duty, movers hired by the Navy came to my apartment in Waipahu and packed up my furniture and other personal goods for shipment to my parent's home in Arcata. I turned in my Vega at the Port of Honolulu for shipment to Oakland and moved into a room at the Waikiki Sheraton for a few days. I then flew to San Francisco and moved to a motel in Alameda California. When my car arrived, drove up to Arcata and spent a week with my parents. I then drove down to Yosemite to spend a week camping there. At the end of that trip, I returned to Alameda and started shopping for a sailboat. I purchased a 24-foot Islander Bahama sloop for about $4500 and spent the next month refitting the boat for single-handed cruising. I added a depth sounder built from a Heath Kit and a wind-vane self-steering system. I named the boat Rubicon in honor of the French code phrase "Situation Rubicon Franchi."

While I was living in the Alameda motel, and preparing Rubicon, I received a letter from my parents with an enclosure from the Navy. I was asked to report to the Naval Station at Treasure Island, between San Francisco and Oakland, to be awarded the Joint Service Commendation Medal. I had been awarded this medal for my contributions to Project Hula Hoop aboard the USNS Wheeling the previous summer. The medal had been awarded in December of 1973 but had wandered through the files at NAVCOMMSTA and CINCPACFLT for several months while I was at sea on Pony Express operations. I showed up in my civvies and was handed the award certificate with no ceremony other than a congratulatory handshake. Had the award been issued for the 1972 operation, it might have tipped the balance toward my retention on active duty.

I spent the rest of the summer sailing around San Francisco Bay, as far north as Bodega Bay and east into the Sacramento River delta. I had considered sailing the Rubicon north to Puget Sound, but the difficulties of the trip to Bodega Bay and the lack of accessible ports on the way north convinced me that I didn't have the right boat for that trip. That decision was confirmed when I got caught in fog and an outgoing tide returning to San Francisco from Bodega Bay. I was awake for almost 30 hours as I struggled through the fog and adverse currents at the Golden Gate. I finally arrived at Angel Island and dropped anchor at about 2AM.

I finished out the summer in the Bay area and moved north to Corvallis to continue working toward my M Sc. degree in Chemical Oceanography. I arranged for Rubicon to be sold by a broker and eventually recovered about $3500 of the $6500 I had spent on the purchase, refit, and expenses for the summer.

I moved back to Corvallis in late September of 1974 and re-enrolled in the School of Oceanography. I had several thousand dollars in the bank and would get $270 per month of GI Bill educational benefits. Shortly after I started school, I was awarded a Graduate Research Assistant position which also paid a stipend of several hundred dollars per month and included a tuition waiver. As a result, I had an income greater than my expenses during graduate school. This situation was even better (and not just financially) after my girlfriend, also a graduate student, moved in with me and shared the rent on our apartment.

The Ultra Secret

In late 1974, English author Frederic Winterbotham published "The Ultra Secret" which described the extent to which Allied decryption of the German Enigma ciphers had influenced World War II. At the time, the Enigma machines and their derivatives were still widely used by many European governments. There was great consternation in the cryptologic community at the publication of this book and the later descriptions of US decoding of Japanese communications. The Enigma machines and their use were well described in the classified cryptology course I had taken, and I was very surprised that the secret information was now public.

In the years since 1974, I have continued to read what I can find about cryptology and the Naval Security Group. Although I have worked on many projects for the Office of Naval Research and other DOD organizations, none required a security clearance. Living outside the cryptologic community is both a blessing and a curse. The curse is that you no longer have "insider information" about technology and US adversaries. The blessing is that you are free to discuss your work with anyone who will listen.

US law concerning classified information mandates that almost all classified information be declassified at most 50 years after its origination.

That means that the operations I have described should be declassified no later than 2024. I have filed several Freedom of Information Act (FOIA) requests for information about the Pony Express operations in which I participated but have received no response at this time (December 2017). Since the NSA has added an exhibit on telemetry intercept to the National Cryptologic Museum, I don't think this memoir includes anything that could not be in that display. After 45 years, I'm not even going to guess at telemetry frequencies or modes.

Chapter 20 Join the Navy, See the World

During my five years in the Navy, I managed to see a lot of the world on Uncle Sam's travel voucher. Transfers between duty stations in the US and abroad were made on Navy travel orders. There were several trips back to visit my parents in Arcata between duty stations that were also paid for by the Navy. When I spent a month in Europe between Morocco and Newport, Rhode Island, I had to pay for my own EurailPass to travel within Europe, but the ticket from Paris to New York was paid for by the Navy. The map below shows the places I visited or lived in. The locations are in chronological order. Some duplicate travels are omitted, as are a few trips from Hawaii back to California on leave. There was also a week of Holiday leave at Christmas time in the middle of my class at OCS in Newport. I can't remember whether I paid for that trip or the Navy paid.

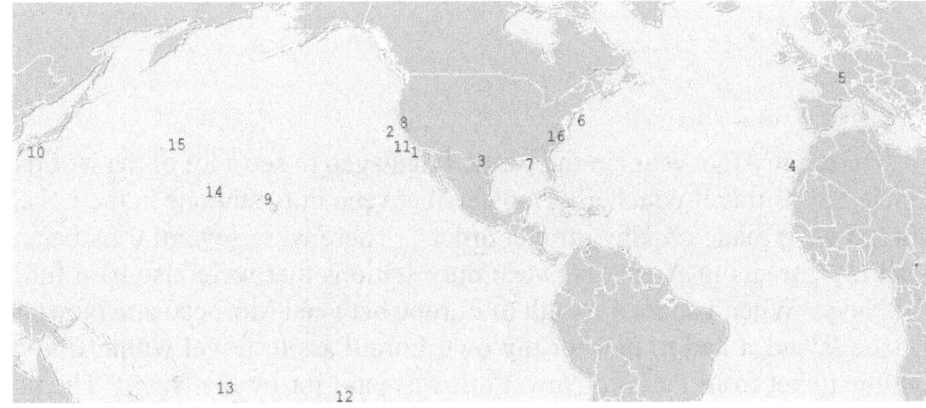

Figure 32 My travels in the Navy

1. San Diego, California	Boot Camp
2. Monterey, California	DLIWC French Course
3. San Angelo, Texas	RT Operator School
4. Sidi Yahia, Morocco	NavCommSta
5. Western Europe	Leave before OCS
6. Newport, Rhode Island	Naval Officer Candidate School
7. Pensacola, Florida	NSG Officer Orientation
8. Oakland, California	End of Cross-Country Drive
9. Wahiawa, Hawaii	NavCommSta Honolulu
10. Yokosuka. Japan	Canceled Pony Express
11. Pt. Mugu NAS	Board USNS Wheeling
12. Mururoa, French Polynesia	Hula Hoop Missions
13. Apia, American Samoa	Resupply During Hula Hoop
14. Midway Island	Pony Express USS McMorris
15. North Pacific Dateline	Pony Express Missions
16. Washington DC	Conference at DNA

Appendix A. Wullenwebers, SIGINT and HFDF

The text in this appendix comes from an excellent document generated by the National Park Service Historic American Buildings Survey. It was written to preserve the history of the FRD-10 HFDF antenna system and the Naval Communications Station at Wahiawa Hawaii, where I was first assigned upon arriving in Hawaii. The document has been converted from a PDF with Optical Character Recognition (OCR) software and reformatted to fit this publication.

U.S. NAVAL BASE, PEARL HARBOR, NAVAL RADIO STATION, AN/FRD-10 CIRCULARLY DISPOSED ANTENNA ARRAY (Naval Computer & Telecommunications Area Master Station, AN/FRD-10 Circularly Disposed Antenna Array)

(Pacific NCTAMS PAC, Facility 314) Wahiawa Honolulu County Hawaii

HABS No. HI-522-B

PHOTOGRAPHS WRITTEN HISTORICAL AND DESCRIPTIVE DATA

HISTORIC AMERICAN BUILDINGS SURVEY U.S. Department of the Interior National Park Service Oakland, California

Location: Wahiawa vicinity City and County of Honolulu, Hawaii

USGS 7.5 minute series topographic map, Hauula, HI, 1992. Universal Transverse Mercator (UTM) coordinates (for the center of the area enclosed by this circular facility): 04.602400.2380260.

Present Owner: United States Navy

Present Occupant: United States Navy

Present Use: Decommissioned

Significance:

The AN/FRD-10 Circularly Disposed Antenna Array (CDAA) at NCTAMS (Facility 314) was a part of the United States' Cold War efforts to gather foreign intelligence information. Along with fourteen other FRD-10 CDAAs worldwide, it was a part of the Naval Security Group's Classic Bullseye network, a program for strategic signals

intelligence (SIGINT) collection and transmitter locating. This CDAA technology, designed by the Naval Research Laboratory and deployed as the FRD-10, was a radical improvement in the performance of high-frequency direction finding. Its design is the Navy's adaptation of an antenna system using monopole and dipole elements uniformly spaced outside the rings of reflector screens. Thus, the system is able to intercept and detect the direction of high-frequency radio transmissions covering 360 degrees.
Date: 2007
Prepared By: Dee Ruzicka Architectural Historian Mason Architects, Inc. 119 Merchant Street, Suite 501 Honolulu, HI 96813

Signals Intelligence Collection

Throughout the post World War II years, advancing technology helped to shift the means of intelligence gathering by the United States, from networks of agents operating on the ground in foreign lands to electronic and over flying systems that could gather data from much greater distances Although human intelligence gathering efforts continued, technical collection of information came to be relied on more and more. An important time for this was the 1950s as the United States began to build a network of signals intelligence (SIGINT) stations to collect information, and also to experiment with aerial reconnaissance over the Soviet Union (Richelson 1995, 256). SIGINT, which is often referred to as "reading someone else's mail," can be divided into two categories. The first, Communications Intelligence (COMINT) is the covert interception of foreign communications. This involves regular message traffic: voice, radio-telephone, facsimile, or Morse code, either transmitted in the clear or encrypted. The second, Electronics Intelligence (ELINT) refers to non-communications signals, such as the emanations from foreign radar, and signals sent back from missiles or satellites that indicate performance and operation during a flight. The FRD-10 CDAA (Facility 314) at NCTAMS Wahiawa and other CDAAs of the same design were used for gathering COMINT signals intelligence.

In 1952 an antenna system was activated in Scotland for intercepting Soviet transmissions, and there the performance of various configurations was evaluated In 1953 efforts were made using that system to intercept radio transmission from military and commercial traffic near Murmansk by the Air Force Security Service's 37th Radio Squadron Mobile, known as USA-55. By 1955, USA-55 was intercepting signals concerning the new Soviet radar systems that were replacing older Lend-Lease units. Turkey was another area that was especially conducive to monitoring the Soviet Union By the mid 1950s Karamursel in Turkey was selected for the location of a mobile radio squadron which would prove valuable in gathering information on the Soviet Naval Training ground in the Black Sea and launches of the Soviet missile program.

Over flights of Soviet territory to collect intelligence also began during this period. First flights were in 1951 by RB-45 jet aircraft over the Soviet

Union at Sakhalin Island. By 1954 over flights of the Murmansk and Kola Peninsula areas had been accomplished, as well as over Siberia and Wrangell Island. Losses of airmen from flights over the Soviet Union during this period were significant. In October 1952, an RB-29 with a crew of eight was lost in a flight originating in Hokkaido. In July 1953 only one airman from a crew of nine was rescued when an RB-50 was shot down by Soviet MiG-15s over the Sea of Japan. Before the end of 1955 there were five additional incidents which claimed the lives of thirteen crewmen. On November 24, 1954 President Eisenhower approved a program to build thirty high-performance aircraft for over-flight intelligence gathering. These aircraft would become the U-2 reconnaissance plane, codenamed IDEALIST, which flew its first Soviet over flight on July 4, 1956. The aircraft was designed totake photographs while flying at altitudes above 68,000 feet, then thought to make it immune to air-defense missiles and interceptors.

Although the Soviets launched the first earth-orbiting satellite (Sputnik I in 1957), it was the United States that moved ahead in the use of that technology for photo-reconnaissance. On August 19, 1960, less than four months after Francis Gary Powers' U-2 was brought down when flying over the Soviet Union, the United States recovered its first reconnaissance satellite payload (Project CORONA) which contained photos of a Soviet air base Later CORONA missions in 1960 returned with photos showing that the "missile gap" between the US and Soviet ICBM inventory did not exist. The first successful Soviet photo-reconnaissance satellite was launched in early 1962. By 1964, the year after the FRD-10 CDAA was constructed at Wahiawa, US satellite reconnaissance (codenamed KEYHOLE) had returned photos with a resolution of eighteen inches (Richelson 1995, 301). Also by the end of that year, the existence of the SR-71 reconnaissance aircraft (codenamed OXCART) was made public by President Johnson. This plane, capable of flying at 92,500 feet at a speed of mach 3.5, would make its first operational flight in May 1967 over North Vietnam.

Other SIGINT schemes from the period range from the mundane, such as the use of small, slow ships, or spy trawlers, which patrolled the Soviet coast, to the exotic, like Project FLOWER GARDEN which collected Soviet signals from radar installations after they had bounced off the moon (Richelson 1995, 305). Amid this carnival of moonbounce, satellites, spy trawlers, and mach 3 aircraft, the United States also pursued the strategy of making more effective antennas to gather SIGINT, such as the high frequency FLR-92 and FRD-10.

High Frequency High-frequency (HF) radio waves were the most common system of telecommunication before the early 1960s. In a curious twist of nomenclature, a high-frequency (HF) signal is actually considered a part of the low-frequency band (which includes Extremely Low Frequency or ELF, Very Low Frequency or VF, Low Frequency or LF, as well as High Frequency or HF). "Messages transmitted at lower frequencies (ELF, VLF, LF, HF) travel for long distances since they bounce off the atmosphere and will come down in locations far from the transmitting and intended receiving locations. In contrast, data sent at higher frequencies will 'leak' through the atmosphere and out into space" (Richelson 1999, 183). HF transmitters and receivers were especially popular with the military and for diplomatic communications, such as between an embassy and its mother country HF signals are useful for these types of communications because of their ability to bounce off the ionosphere, the upper region of the earth's atmosphere which begins about 50 miles above its surface. This means that they are able to be transmitted to receivers that are over the horizon, behind the curvature of the earth from the emitter. A powerful HF signal has the ability to travel around the entire planet for reception, which makes it a good choice for global endeavors and also makes it extremely vulnerable to interception (Campbell 1999, 8). Another name for high frequency radio (operating between the frequencies of 2,310 kHz and 30 MHz [30,000 kHz]) is shortwave. This is because there is an inverse relationship between frequency and wavelength; high frequencies are associated with short wavelengths.

Evolution of the FRD-10

Early work on circularly disposed antenna array (CDAA) systems was undertaken by the German navy's signal intelligence research and development center early in World War II. It was during this time that CDAA was given the name, Wullenweber, or Wullenweber Antenna. Jurgen Wullenweber was the mayor of Lubeck, Germany from 1533 to 1537 He was an opponent of injustice and a supporter of the Protestant cause who became a legendary figure. Considered a martyr after he was killed in 1537, his name was chosen as a cover for the German CDAA program of World War II. After the war, some of the German CDAA technology was appropriated by the Soviets, who deployed 20 CDAAs during the post-war period when the United States military showed little interest in them. Two CDAAs were built by the Germans; the one in Denmark was destroyed after the war The other, at Langenargen, Germany, was dismantled, re-erected at the University of Illinois, and studied by Professor Edgar Hayden of the Electrical Engineering Department. Two of the CDAA antenna designs which resulted from Hayden's research were the AN/FLR-9 system that was used by the U.S. Army and Air Force, and the AN/FRD-10 (Facility 314) system that was used by the Navy.

In 1994, an aging Wullenweber direction finder that was used by the University during the 1950s and 60s for development of the U.S. Navy CDAA system, was still located at the University of Illinois facility near Bondville, Illinois. This antenna (unknown if it was the original CDAA re-erected after WWII) had been abandoned around 1980 by the University after the completion of their developmental work (Swenson 1994).

During the 1950s and 1960s the Naval Research Laboratory (NRL) worked on ways of improving the capabilities and performance of high-frequency direction-finding (HFDF) equipment One product of this research was the refinement of circularly disposed wide- aperture direction finding arrays, or circularly disposed antenna array (CDAA). These circular arrangements of concentric antennas markedly boosted the ability of HFDF systems to locate and collect signals (De Young 1998, 28) Before the circular concept was built, the NRL experimented with a wide-aperture linear antenna 1100' long at Fox Ferry, Maryland around 1952. This device operated well only within the narrow range of its orientation, but was able to determine bearings of transmissions up to 1200 nautical miles away to an accuracy of 0.25 degree (Gebhard 1979, 311). By 1957 the NRL had modified this linear-type antenna into a circular version about 400' in diameter that was built at Hybla Valley, Virginia. This direction finder was used to determine the orbit of the first man-made earth-orbiting satellite, Sputnik I, launched on October 4, 1957 (Gebhard 1979, 313).

Another product of the NRL work on HFDF equipment was the development of retrospective direction finding, or the ability to locate the position of a transmitter after the transmission had been completed. This was accomplished by the use of a high-speed recording device that was coupled with the CDAA system (Bamford 2001, and DeYoung 1998, 28). The first use of this technology was during Project Boresight in 1960, which used an early CDAA system, the AN/FLR-7. This was a 1300' diameter circular array, 200' high. Project Boresight was initiated after November 1960 when National Security Agency (NSA) operators, who had been routinely intercepting daily HF signals from Soviet submarines suddenly noticed no signals. The missing signals proved a mystery until about a month later when it was determined that the Soviets were compressing the signals and shooting them out in a fast burst that was too quick to be pinpointed. The rapidly developed retrospective direction finding system was coupled with the FLR-7 to accurately locate the source of the signal and record it for de-coding (Bamford 2001).

Even before the FLR-7 achieved notoriety in Project Boresight, a successor CDAA, the FLR-9 was under development The FLR-9 was used by the Air Force and the Army The first contract to construct two 1200'-diameter FLR-9 Air Force antennas was awarded in 1959 to Sylvania. These were completed in 1962 at San Vito, Italy and another at RAF Chicksands, Great Britain. The FLR-9 antenna had three concentric antenna rings and such installations, as well as the later FRD-10 CDAA, were often referred to as "elephant cages." The outer two rings of antenna on the FLR-9, placed at diameters of 1198' and 1116', were assisted by a reflecting screen of vertical wires that was placed at a diameter of 1076', inside the middle antenna ring. The third ring of antenna, the innermost ring, was placed at a diameter of 334' and had its reflecting screen at a diameter of 314'. The entire array was built on a ground screen about 1443' in diameter. At the center of the array was a building which housed only the system electronics; operators were located in a separate building. The FLR-9 was considered the most advanced CDAA of its period, capable of intercepting and getting a bearing on "as many directions and on as many frequencies as may be desired" (Campbell 1999, 9) Additional FLR-9 antenna were constructed at; Misawa Air Base, Japan; Clark Air Base, Philippine Islands; Elmendorf Air Base, Alaska; Karamursel Air Station, Turkey; Augsburg, Germany; and Udon Thani, Thailand. During its operational history the world-wide network of FLR-9 CDAA was known as program "Iron Horse" and was used for signals intelligence gathering, or eavesdropping, on foreign government communications The Misawa station monitored "diplomatic communications, communications involving Soviet or Chinese strategic nuclear forces, and Soviet satellite communications (Richelson 1987, 94). Operation Ladylove, using the FLR-9 at Misawa intercepted communications signals from Soviet geosynchronous satellite systems: Molniya, Raduga, Gorizont (Richelson 1989, 183, and Elliston 2004). The FLR-9 at Misawa Japan can "pick up a Russian television broadcast in Sakhalin or an exchange of insults between Chinese and Soviet soldiers on the Sino-Soviet border" (Beech 1980, 17.) By 1996 all but three FLR-9 CDAAs had been dismantled, Elmendorf and Misawa were still standing, and Augsburg was to be turned over to the German government in 1998.

The Navy's version of the CDAA, the FRD-10, was slightly smaller than the FLR-7 and FLR-9. Its Navy predecessor, the AN/GRD-63, was a small system of only eight monopole antenna elements that was developed in 1951 (Gebhard 1979, 310) The network of FRD-10s was developed to plot the locations of Soviet submarines and other HF radio transmitters and was managed by the Naval Security Group (NSG). Unlike the Army/Air Force FLR-9, the FRD-10 had only 2 rings of antenna, located at about 870' and 780' diameters. Each ring was backed by a reflecting screen of vertical wires that was placed inside its diameter. While the third ring of dipole antenna elements on the FLR-9 gave the unit a wider frequency, its horizontal polarization interfered with its omnidirectional capabilities, and was difficult to operate.

The FRD-10 was an improvement by offering a "truly omnidirectional pattern" (Cummings 2000, [4]). The FRD-10 was designed as an outward-looking array, with its high-band antenna elements on the outside, backed by a dedicated reflector screen. This is in contrast to the inward looking FLR-9 which had its high-band elements on the inside of its circular structure

Construction of the Navy's FRD-10 CDAAs was undertaken between 1962 and 1964. During that time a total of 16 were built, one each at the following 14 locations: Adak, AK; Marietta, WA; Skaggs Island, CA; Imperial Beach, CA; Wahiawa, HI (NCTAMS); Guam, Mariana Islands; Hanza, Okinawa, Japan; Winter Harbor, ME; Northwest, VA; Homestead, FL; Sabana Seca, PR; Galeta Island, Panama; Edzell, Scotland; and Rota, Spain. Two additional FRD-10 CDAAs were built in a side-by-side configuration at Sugar Grove, VA. These were used for ship-to- shore communications All of these CDAA have been removed except for Waihiawa and Imperial Beach (both slated for removal in 2007), Galeta

Island (equipment removed and turned over to the Panamanian Government), and Guam (abandoned). The FRD-10 antenna system was used at Homestead AFB in Florida to monitor "Cuban military communications as well as Soviet activity in Cuba" and "all communications involving Cuba and Soviet air activity originating or destined for Cuba." (Richelson 1989, 186). Also FRD-10 at Guantanamo Bay and at Sabana Seca, Puerto Rico, "intercept Cuban and Soviet military communications in and around Cuba and the Caribbean Basin" and "target internationally leased carrier (e.g. INTELSAT) and diplomatic communications for all of Central and South America" (Richelson 1989, 186) The FRD-10 was built in two different variations After the first one was constructed at Hanza, Okinawa, there were some dimensional changes that were incorporated in later units, such as the diameters, heights, spacing, and number of antenna elements (Cummings 2000, [5]).

The FRD-10 CDAA has a normal range of 3200 nautical miles. This distance corresponds to one or two bounces of a HF signal off the ionosphere. While it is often possible to monitor multi- bounce signals, the direction-finding accuracy is generally compromised. However, under good conditions for the propagation of signals, a transmission can be monitored from all around the earth (Granite Island 2006, 2) CDAA antennas are designed for optimal performance with incoming wave angle of 10-30 degrees or 45-60 degrees. Although these antennas are omni directional, their locations were likely chosen to place it "purposeful[ly] relative to the intended spying target(s)" (Cummings 2000, [2]).

Classic Bullseye and the NSG

The AN/FRD-10 CDAA at NCTAMS (Facility 314) was built in 1963 as part of the network of 14 land-based, high-frequency direction finders of the same design that were used in the Classic Bullseye program. This Department of Defense program gathered strategic intelligence and determined the location of transmitters. Classic Bullseye was managed by the Naval Security Group (NSG) Command. The High-Frequency Direction-Finding (HFDF) FRD-10s were used to intercept and locate voice transmissions and message traffic that were broadcast on short-wave (HF) channels (Pike 2001). In the late 1960s, an FRD-10 CDAA was constructed at Canadian Forces Station Masset, located on the north coast of Graham Island in British Columbia's Queen Charlotte Islands. This station officially opened in February 1970, and became a station of the United States' Classic Bullseye program. By 1978 a permanent detachment of about ten NSG personnel were at Masset as part of a joint Canadian Forces/U.S. Navy exchange program. Canadian technicians would also train at the NSG's school in Pensacola Florida (Proc 2006 [8]). The station's official role in Classic Bullseye was to "a/ participate in the Canada/United States HF DF search and rescue net, b/ support in the collection of data for research into basic problems on ship/shore and shore/ship communications, c/ HF DF assistance to search and rescue operations."

The station was "believed to listen to the Soviet naval base at Petropavlovsk and to the Vladivostok headquarters of the giant Soviet Pacific ship and submarine fleet" (Granite Island Group 2006). Also under Classic Bullseye, the Masset station eavesdropped on Soviet spy trawlers.

The Naval Security Group was the portion of the Navy which conducted radio intercept (intelligence gathering) and crypto analysis between its formation in 1935 and 2005, the year it was disestablished. Some of the first radio intercept activities undertaken by the Navy date from 1923 in the Pacific when the Office of Naval Intelligence directed all ships in the Asiatic Fleet to forward Japanese and commercial coded traffic which was intercepted (Grobmeier 2006). By the following year Japanese radio messages were being intercepted by the Navy in Shanghai, China. In March 1935 the forerunner of the NSG was formed, the Communication Security Group. This entity dealt with the Navy's radio intelligence and cryptology operations. Over the next 20 years the Communication Security Group expanded, and in 1955 the title Naval Security Group was applied. The mission of the NSG was to "perform cryptologic and related functions." It also "manages the Navy High Frequency Direction Finding System and associated communications support" (Pike 1997). The 2005 disestablishment of the NSG was part of a reorganization, and the commands formerly under the NSG were renamed Navy Information Operations Command (NIOC), whose mission continued to be to "conduct information operations and to provide cryptologic and related intelligence information" (NIOC 2006).

Operation of the FRD-10

The two rings of antenna in the FRD-10 CDAA are set up to each operate optimally in a different high-frequency (HF) wavelength range. The outer ring of 120 sleeved monopole antennas was designed to receive shorter HF wavelengths from 8-30 MHz and the inner ring of 40 folded dipole antennas the longer HF wavelengths between 2-8 MHz (Proc 2006, [12]). A folded dipole antenna gives enhanced performance over a standard dipole by allowing an increase in feed impedance and wider bandwidth capabilities The two screens of vertical wires that are suspended from the wood poles serve to shield the antenna elements on one side of the CDAA from interference that would be created by signals crossing from the other side. The wire shield buried in the ground also serves to isolate the incoming signal for better results.

The individual antennas of the ring were sequentially connected to radio receivers in the operations building in the center of the array. A contiguous number of antenna elements which spanned a segment of the circle of the CDAA were usually connected at any given moment. This segment of activated antenna formed a pattern which swept rapidly around the ring, covering all points of the compass. An antenna which sweeps its beam using electronic pulses moving across its face is termed "phased." This is in contrast to a manually swept antenna, such as a rotating satellite dish. By monitoring the signal strength and precise timing as the activated segment moved around the circle, the direction of a radio signal could be very accurately determined Each ring of the antennas was relayed through a goniometer (an instrument that allows a precise determination of the angle of a rotating object or signal). The goniometer switches the antenna elements sequentially alternating connected and disconnected elements to produce a beam that swept around the CDAA and covered all directions. The details of the phasing of the antenna elements and the goniometer circuitry are "presumed to be classified"

The real unique use of the CDAA antenna is that certain elements can be selectively chosen and electronically added or subtracted in phase effectively synthesizing an array that behaves as if the elements were in a linear end-fire array with substantial gain. This permits the construction of electronic "sector" beams that are 30 degrees wide or "monitor" beams that are 15 degrees wide.

The FRD-10 is designed to react to incoming signals as if it were a linear array. This is accomplished with the sophisticated circuitry and relays of the goniometer. As the incoming signal hits the array it reaches a single antenna element first and then subsequently triggers pairs of elements sequentially moving outward as the wave moves over the array. The signals are evaluated at the goniometer and combined with a delay so that each element's signal arrives simultaneously at the receiver, performing as if the circular array were linear and oriented almost directly at the source of the signal. "The goniometer permits the user to in effect electrically steer the phasing/delay lines by selecting a given arc of [antenna elements] on demand. The resulting pattern is very narrow and the gain very high. It shines in allowing the listener to electronically steer the beam and change the pattern shape at will" (Cummings 2000, [7]).

The various FRD-10 sites around the globe were in contact with one another which enabled triangulation of the source of the signal by comparing bearings from two or more FRD-10s. "The FRD-10 provided a near instantaneous bearing of any signal that appeared on the radio spectrum for even a fraction of a second.

When combined with the information from other FRD-10 sites operating in real time, a bearing could be obtained immediately and it would be virtually impossible to hide any HF transmissions (Proc 2006, [12]).

For the Navy, the FRD-10 CDAAs were a big improvement over their earlier HF direction finding unit, the GRD-6 from 1951. They allowed transmissions to be recorded for subsequent direction finding. The bearings obtained were up to four times as accurate as previous antennas, with results of better than 0.5 degree. Signal amplitude (gain) was four times greater, and the FRD-10 was better able to filter out interfering signals and noise (Proc 2006, [15]) "The improvement expected as a result of deploying the FRD-10s was a combination of more accurate and reliable fixes, producing reduced search areas in ocean areas of prime responsibility so fresh in time, as to enable maritime commanders to deploy their forces more economically and with much greater prospect of making contact with the target than is now the case.

History of Communication and Signals Intelligence at NCTAMS

The development of the Naval Computer and Telecommunications Area Master Station, Pacific (NCTAMS PAC) at Wahiawa was begun in 1940 as a temporary radio transmitting station and radio direction finder The need to expand the Navy's receiving capabilities on Oahu in anticipation of war led to the incorporation of receiving facilities at the Wahiawa installation. Prior to the development of this base, the main Oahu communication stations for the Navy were the Naval Radio Station on the west shore at Lualualei, and a receiving station at Wailupe on the south shore. The Wailupe station was the site of the intercept facility for Station Hypo, which had been intercepting Japanese communications in an attempt to break their code). The Lualualei and Wailupe stations functions were consolidated at the Wahiawa location by the end of 1941 (Global Security) and in 1942 the designation for this installation was "Naval Radio Station for the 14th Naval District" (14th Naval District 1942).

During World War II, this base was "the principal radio receiving station in the Hawaiian Islands: with its associated transmitting stations, it constitutes the main link in the naval-communication chain between Washington and the Pacific combat area" (CPNAB A-856) The facility sent messages and also had units responsible for cryptographic security, message traffic control, and message traffic analysis.

After the war, the Naval Research Laboratory (NRL) continued work which was carried out by the Army's Signal Research and Development Laboratory in 1946; this consisted of reflecting radar signals off the moon. The NRL conducted investigation into the possibility of using signals bounced off the moon for relaying military communications In early 1956, NCTAMS participated in relaying communications between Wahiawa and Washington D.C. via the moonbounce of signals. That year NCTAMS became a station (along with the radar site at Opana on Oahu's north coast) for the Navy's Communication Moon Relay (CMR) system which sent teletype and facsimile transmissions between Washington D.C. and Hawaii The equipment consisted of two 84' diameter dish antennas, one each for transmitting and receiving. The NCTAMS antenna was located north of Polaris Drive (Helber Hastert & Fee Planners 2006, 5.2.3). The CMR system became operational in January 1960 when a message was sent from Washington to the Commander of the Pacific Fleet. The CMR system operated between Hawaii and Washington until the mid 1960s. It was reliable, its main limitation being the position of the moon. Operators manned the system for a period of four to eight hours daily, the time from moonrise in Hawaii to moonset in Washington.

NCTAMS also became the site of a satellite tracking station (station number 100) in the Navy's Tranet system, which determined orbiting position by detecting signals transmitted from the satellites. The Tranet system had seventeen stations in the early 1960s. When the Navy's Transit system (a fore runner of the Global Positioning System) became operational in 1967, the Tranet station at NCTAMS was switched to a tracking station for it. Transit was a satellite navigation system that employed "four or five" satellites and ultimately gave users accuracy of about 25 meters by determining position in relation to any two of the satellites (Federici 1997, 2.2). The Transit system was ended in December 1996.

In 1964 the Navy established the Technical Research Ship Special Communication System (TRSSCOM) This was the first ship-to-shore satellite communication system and was designed to support Navy SIGINT surveillance ships that were operating in the field gathering intelligence. The TRSSCOM name was chosen to remain in keeping with the ships' cover story that they were involved, not in surveillance, but in technical research (Federici 1997, 2.3.1). When the CMR moonbounce link between Hawaii and Washington was shutdown, the CMR facility antennas at NCTAMS were cycled to TRSSCOM (Federici 1997, 1.6.2.1), going operational in 1964. TRSSCOM was ended in the fall of 1969.

In 1967 President Johnson denied a Soviet claim that two of their ships had been damaged when the United States bombed Haiphong Harbor in Vietnam. Upon having to eat crow when the Soviets produced photos of the damage, the President mandated that in the future he was to personally view military photos from Vietnam. He also wanted the photos sooner than they could be courier-delivered from the field. The response to his order was an operation, called Compass Link, which transmitted photos from Vietnam via satellite to NCTAMS and then on via satellite to an NRL 60' dish antenna at Waldorf, Maryland. Compass Link was used until the end of the war in Vietnam.

NCTAMS is a transmitting and receiving station for the Fleet Satellite Communication (FLTSATCOM) System. This was a project undertaken with the Air Force and begun in 1971 that allowed almost worldwide communication via a system of four geosynchronous satellites. The involvement of NCTAMS, along with the other stations for FLTSATCOM (Washington D.C., Norfolk, VA, and Finegayan, Guam), has made it "one of the most vital fleet communications centers in the Navy" (Helber Hastert & Fee Planners 2006, 5.2.4) On February 18, 1977 NCTAMS was officially designated the site for a Super High Frequency Satellite Facility, the largest facility of this kind. "In 1980, Wahiawa was the largest communication facility in the world" (ibid.).

Ground-Based Signals Intelligence at the End of the Cold War

The end of the Cold War and the dissolution of the Soviet Union in 1991 brought a series of dramatic cutbacks in Signals Intelligence networks as threats of a conflict between the Soviet Union and the United States vanished. During the late 1980s and the 1990s a number of major United States SIGINT stations that previously monitored the Soviet Union were closed: in Italy (San Vito), Germany (Augsburg and Berlin), United Kingdom (Chicksands), and Turkey (Sinop). By 1999 "the National Security Agency [had] established three regional SIGINT Operational Centers to received data from manned and unmanned SIGINT sites in particular regions" (Richelson 1999, 197-8) A center at Lackland, Texas was focused on Latin America, Fort Gordon, Georgia had a station concentrating on Europe and the Middle East, and a station at Kunia, Hawaii was focused on Asia.

In the early 1990s a tactical HFDF, the AN/TSQ-1644 (Code named DRAGONFIX) was being experimented with. This system was field-transportable and operated within a frequency of 1.6 to 30 MHz. It was able to determine the angle at which signals bounced off the ionosphere to arrive at the receiver. This enabled a single TSQ-164 receiver to not only get a bearing, but to also calculate the position of the transmitter without any triangulation (Proc 2006, [16]).

Beginning in the mid-1990s the NSG, noting the absence of Soviet targets and wanting to cut costs and change the focus of its SIGINT collection, began closing FRD-10 sites. As of 1999 the FRD-10 at NCTAMS Wahiawa was still being used to peer into the communications of various entities around the Pacific region The most specific information available on this indicates that the FRD-10 was used "to monitor naval traffic around the Hawaiian Islands as well as to collect international leased carrier (e.g. INTELSAT) and other communications for the Pacific Region" (Richelson 1999, 200-201). This seems to indicate that the elephant cage was tuned to the sky and listening to any targets deemed interesting by its commanders.

At this date the FRD-10s at Imperial Beach CA and in Guam were also operating. They have since been closed.

Undoubtedly, since the September 11, 2001 terrorist attack on the World Trade Center and the Pentagon, listening posts have gained importance and most likely increased in number and sophistication. The FRD-10 CDAA at NCTAMS Wahiawa ceased listening in August 2004; it can only be assumed the closure occurred because there was a better way to do it.

Appendix B. Declassified Hula Hoop Documents

Here are some of the declassified documents pertaining to Hula Hoop operations in 1972 through 1974. Most of the text here has been extracted from much larger PDF documents, converted with Optical Character Recognition (OCR) software, then edited and formatted to fit this document.

CINCPAC Command History 1973 Excerpt.

Note that paragraph classification indicators have been removed and the OCR text has been reformatted.

HULA HOOP

French nuclear testing in the vicinity of the Tuamotu Archipelago had become a regular summer event in the PACOM. The long-range program for U.S. reconnaissance of those tests was called NICE DOG. Each season's tests, however, also had a nickname and in 1973 the surface aspects of the tests were called HULA HOOP. On 20 March the JCS informed CINCPAC that the tests were expected to begin about 20 July, providing an opportunity to accumulate data not otherwise available. The JCS requested that CINCPAC provide administrative support to the Defense Nuclear Agency's (DNA) Joint Project Office (DJPO) established for the HULA HOOP operations and provide operational control of and support for the surface collection platform, USNS WHEELING (T-AGM-8), which was called POCK MARK. Other data collection platforms assigned to HULA HOOP were a KC-135 and an NC-135, but these were under the operational control of CINCSAC.

In anticipation of the testing, CINCPAC tasked CINCPACFLT with operational control and a support plan for WHEELING, a search and rescue plan for the SAC aircraft utilizing WHEELING, a plan for utilizing WHEELING as a communications relay platform for the SAC aircraft, and a joint communications plan to support all units involved in HULA HOOP. CINCPAC requested that CINCPACAF pro- vide administrative support similar to that provided for the 1972 program (DIAL FLOWER) to the DJPO.

CINCPAC subsequently approved a request by the Commander of the Pacific Missile Range at Point Mugu, California to present a HULA HOOP briefing to the civilian crew of USNS WHEELING. The briefing, which was presented early in June, covered the following thoughts:

> USNS WHEELING has been assigned by the Joint Chiefs of Staff as an essential observation platform to acquire important data from foreign atmospheric nuclear tests to be conducted this spring in the South Pacific Ocean area. WHEELING has been selected for this task due to the modern equipment installed and availability of space for additional instrumentation. A deployment of about 60 days is presently planned with an in-port period at Honolulu included.
>
> While the presence of WHEELING in the test area will be known by the foreign power conducting the tests, the U.S. Government will make no announcement of the ship's position or mission. For this reason, the mission of WHEELING must be considered as classified information, and not be discussed at any time while off ship.
>
> Positioning of WHEELING to acquire the needed data has been carefully determined-with the safety of embarked personnel the primary concern. Blast, heat and radiation were each considered in detail to assure that the safety of personnel will not be jeopardized. Knowledge of each effect has been acquired over many years from U.S. tests of much larger magnitude than WHEELING is expected to observe. At the ship's position, blast effects will be detected as a loud report accompanied by a slight jolt and followed by a noticeable wind (maximum 18 knots).

> Thermal output will be detected as a slight warming of the skin by topside personnel. No nuclear radiation will be experienced because of the ship's upwind position.'

On 18 May the JCS advised that the Director of the DNA had been informed that fleet ship assets were not available to support augmentation of the basic HULA HOOP program with a radar transmission experiment. The use of USNS CORPUS CHRISTI BAY (T-ARVH-1) was offered as a suitable-replacement and the DNA Director determined that it was acceptable to meet surface ship support requirements. The JCS requested that fleet SH-3A helicopter assets and required maintenance support be provided to support the radar transmission experiment. Accordingly, on 19 May CINCPAC directed CINCPACFLT to provide these helicopters to assure continuous availability of two operational units during the test period.2 On 30 June CINCPAC informed the JCS that the HULA HOOP Technical Coordination Plan for 1973 had been reviewed as requested. The technical content and platform positioning graphics contained in the plan were considered operationally feasible and acceptable. The ship's tracks were listed.

On 11 July the JCS issued an alert an execute message. They advised that approval had been granted for NICE DOG/HULA HOOP 1973 operations. A series of four nuclear test detonations was expected to be conducted by France on or about 20-25 July and 20-25 August at the Mururoa Atoll South Pacific test site.

Data collection tasking was as follows. POCK MARK and USNS CORPUS CHRISTI BAY (nicknamed POT LUCK), with embarked SH-3A helicopters, were to utilize all sectors in the vicinity of the atoll with a closest-point-of-approach of 20 nautical miles to ground zero immediately prior to, during, and immediately after the detonations. The position the ships were to maintain in relationship to one another was outlined. All U.S. monitoring resources were to remain clear of the 12-nautical mile limit of French territorial waters at all times. The SAC aircraft bore the nickname BURNING LIGHT, as they had in previous years.

The JCS requested that contingency withdrawal plans be prepared to meet the circumstance in which POCK MARK and/or POT LUCK were ordered out of their operating areas by the French, or in which the ships were asked to stop radiating.

CINCPAC passed the planning responsibilities to CINCPACFLT; CINCPAC subsequently approved the submitted plans for planning and execution, as appropriate. It was not necessary to execute these withdrawal plans, however.

The first French test was detonated on 21 July, the second on 28 July. All platforms were in position; there were no casualties to personnel or platforms. On 30 July, the JCS authorized that the forces could stand down, at the discretion of CINCPAC, until the next series of tests was expected. On 31 July CINCPAC directed a stand down for other operations as required by CINCPACFLT, but were to assume a 72-hour collection alert status by 16 August.

On 18 August the third test occurred, and another on 24 August. On 28 August, they concluded a test of an air-dropped device. In this case POCK MARK was not on the proper heading for data collection because it was not equipped with air surveillance radar.

On 30 August, the DNA advised that there were indications that the French planned to continue testing in September, possibly to include a high yield nuclear device, and on 1 September the JCS advised that approval had been granted to continue collection operations. The sixth and last detonation of the season was a safety test detonation on 13 September and CINCPAC directed all HULA HOOP forces to return to normal operations on 15 September.

The commander of the task unit (CTU 30.2.6) forwarded his summary report on 23 September. He noted that this was the first time the operations had used two surface vessels and helicopters in close proximity to Mururoa, but that the unit's presence appeared to cause no adverse reactions by the French, possibly because of their being advised in advance of U.S. intentions by the Defense Attaché. He recommended that this procedure be continued in future operations. He concluded:

> "Many elements comprising this task unit were identified and assigned in such late time prior to deployment that no pre-sail opportunity, for operational practice was possible although precise positioning and timing were essential. Operational plans

and procedures were under continuing change and refinement as the mission progressed. The successes recorded in mission accomplishment are considered primarily the result of the professionalism, skill, imagination, and extreme motivation exhibited by numerous personnel embarked, and in the staffs and supporting commands ashore."

SAC Reconnaissance Report FY 74

This excerpt from the Strategic Air Command Reconnaissance Report for FY 74 describes their participation in the Hula Hoop missions. It is important to note that SAC had a lot more problems in the summer of 1974, when there was no NSG Direct Support group aboard the ship at Mururoa (The USNS Huntsville).

BURNING LIGHT

The diversity, responsiveness, and experience of SAC reconnaissance operations and personnel were again demonstrated by the BURNING LIGHT program, conducted to measure the data given off during the test of nuclear devices. Because of the United States-Soviet moratorium on atmospheric testing, SAC had monitored only French tests since it was assigned this mission in 1971. BURNING LIGHT was the airborne portion of a larger nuclear collection effort, which the Defense Nuclear Agency (DNA) nicknamed HULA HOOP in1973 and DICE GAME in1974. Basically, the BURNING LIGHT mission aircraft satisfied a single objective. This was to support the development of a miniaturized, inexpensive, highly sophisticated system for analyzing data from nuclear explosions and to gather information that would improve the United States' ability to predict effects of low-altitude nuclear weapons. A U.S, Navy ship, the Huntsville, also participated in the HULA HOOP and DICE GAME programs. Operating in international waters outside the Pacific test area, the Huntsville monitored the nuclear blasts, and the Defense Nuclear Agency launched drones equipped for nuclear sampling from its deck. The combined program provided a-valuable means of expanding American knowledge about the effect of nuclear weapons for relatively moderate amounts of money, equipment, and manpower.

France conducted its nuclear tests approximately 2,700 nautical miles south, southeast of Hawaii on the Mururoa Atoll in the Tuamotu Archipelago of French Polynesia. These tests normally took place from June to August, a period when winds and other climatic factors were most favorable.

Nearly all French nuclear devices were detonated from balloons, but an occasional one was dropped from aircraft.

FY 74 was a time of extensive nuclear testing by the French. A BURNING LIGHT deployment took place in the summers of 1973 and 1974. The magnitudes of the 10 nuclear explosions occurring during the two summers ranged from .12 to 155 kilotons. In both years SAC deployed reconnaissance crews from the 55th SRW, along with KC-135 tankers and supporting maintenance personnel to Hickam AFB, Hawaii (OL-HB). Upon receiving notice of an impending detonation, the reconnaissance mission aircraft proceeded to the vicinity of the test range and orbited before, during, and after detonation.

In 74 the Air Force Special Weapons Center (AFSWC) of Air Force Systems Command provided the NC--135As, which served as the BURNING LIGHT mission aircraft. This was because SAC's KC-135R aircraft, which had monitored the French tests in FY 73, had been reconfigured for other missions. While one of the R models was responsible for collecting signal intelligence in the Cuban area, the other aircraft could operate as a "special" tanker, having the ability to both onload and offload fuel.

Although a SAC-owned aircraft no longer collected nuclear data, national intelligence users probably did not want to lose the experience and expertise that SAC reconnaissance specialists had acquired during the previous years of BURNING LIGHT operations. SAC was. still required to furnish the task force commander, front end flight crews- for the NG-135A, and the tanker force along with associated maintenance personnel.

The amount of nuclear data collected by the aircraft was sometimes affected by factors over which SAC had no control. The success of any mission depended upon the accuracy of collateral Intelligence sources which furnished the date and the time detonation was anticipated. Last minute postponements and cancellations could limit the amount of collection gathered or prevent it entirely. Since the NC~135A refueled just before entering its orbit area and had only enough fuel to orbit for about 2-1/2 hours before beginning its return flight to Hickam, accurate, on^-the-scene intelligence information was critical to the success of any mission. Also serving aboard the NC-135A were personnel assigned to the Defense Nuclear Agency, the Atomic Energy Commission, the United States Air Force Security Service (USAFSS), and the Air Force Technical Applications Center (AFTAC

For the most part, the 1973 BURNING LIGHT operation proceeded routinely. Two NC-135As, one under the sponsorship of UNA and the other under the Atomic Energy Commission, four- reconnaissance crews, 13 tanker crews, nine KC-135As, and tanker maintenance personnel deployed to Hickam between 12-19 July 1973. All five nuclear tests were monitored between 21 July and 28 August 1973. Both NC-135As were launched for each event, and eight KC-135s refueled them on every mission. Useable intelligence was collected for all but the second detonation which occurred on 28 July 1973. Technical difficulties were believed to have postponed this test until early afternoon (2303Z) but, by this time, the two NC-135As were approximately 1,500 NM north of the test area on their way back to Hickam. The French usually interrupted their test series for about two weeks, probably to evaluate the results of tests already conducted. During this time, the SAC task force had normally redeployed to the CONUS. Not so in 1973. On 30 July 1973, largely to save money and Jet fuel, the JCS directed that SAC's BURNING LIGHT package would remain at Hickam. The resulting savings was approximately $100,000. When French testing resumed in mid-August 1973, both KC-135AS successfully monitored the last three tests that took place on 18,24, and 28 August 1973. The task force began redeploying on 16 September and all personnel, equipment, and aircraft had returned to their appropriate CONUS base by 19 September 1973.

The French conducted no more nuclear tests until June 1974. Through no fault of SAC, the 1974 BURNING LIGHT deployment, from 5 June through 15 August 1974, experienced several problems. Since the French usually detonated their nuclear devices between 7 and 8 a.m. local time, the NC-135A and supporting tankers had to leave Hickam around midnight if the former was to be in the orbit area, more than 2,600 nautical miles away, when detonation occurred. However, noise abatement restrictions in effect at the Hickam/Honolulu International Airport prohibited water augmented takeoffs between 9 p.m. and 7 a.m. This further complicated mission planning. Water augmentation meant that water was injected into the tankers' Jet engines during takeoff to produce additional thrust. The procedure created a great deal of noise, but without it, both the NC-135A and the tanker force had to launch with lighter fuel weights. Headquarters SAC, the Defense Nuclear Agency, and the Joint Chiefs of Staff tried unsuccessfully to persuade the Pacific Air Forces, the command responsible for determining flight procedures at Hickam, to waive the restriction. The reason was that the objections of Hawaiian politicians and environmental groups were simply too strong. What was the impact of the noise abatement policy upon the 1974 BURNING LIGHT reconnaissance program? Essentially, it required a complete realignment of tanker operations. Instead of requiring the usual four tankers and one ground spare KC-135 to support each mission aircraft, two specially configured KC-135s, equipped both to onload and offload fuel in mid-air, five regular KC-135As, and one ground spare were needed for each operational flight.

Each of the five BURNING LIGHT missions flown in 1974 was conducted in the following manner. The single NC-135A launched, followed immediately by three standard KC-135As in a cell. About 20 minutes later a "Christine" tanker and two more KC-135As also launched. The NC-135A and its associated "Christine" tanker were refueled three times each by their associated tanker cells prior to their rendezvous for a final refueling. Just before the mission aircraft entered the collection area, some 3-1/2 hours after takeoff. This procedure allowed the HC-135A to orbit for about 2-1/2 hours before beginning its return flight to Hickam. Total flight time for the mission air- craft was approximately 16-1/2hours, while the duration of the tanker sorties varied from two to 14 hours.

Repeated postponements and cancellations of detonations by the French prevented the collection of as much nuclear data as desired during the 1974 BURNING LIGHT operation. For example, although the first test was conducted on 16 June 1974, the NC-135A did not launch because intelligence sources failed to indicate that an event was to take place. The mission aircraft was in the orbit area four times during the last nine days of June and on 6 July. On each occasion, no detonation occurred in spite of positive indications that balloon launches were imminent. On 7 July 1974, "the NC-135A was again in its orbit area awaiting a predicted event, but fuel limitations forced it to begin the return flight to Hickam before detonation finally took place at 2315Z., much later than usual,

Similar problems continued until mid-August 1974. Fifteen continuous days of bad weather in the test area during July apparently complicated traditional testing patterns. Departures from usual operational procedures and French technical difficulties had turned the Defense Nuclear Agency's decisions to launch the NC-135A largely into "guess work," Continual postponements and late changes in plans had made it impossible for the NC-135A to be in the orbit area for any of the five events that had occurred so far; moreover, funds allocated for the 1974 BURNING LIGHT operation were nearly exhausted. Because of these problems the Defense Nuclear Agency and the Atomic Energy Commission decided Jointly on 16 August 1974 to terminate the airborne, SAC portion of the 1974 nuclear collection program. Although the French made two more detonations during this test series, the BURNING LIGHT task force-began redeploying to the CONUS on I6 August, and the movement was completed on 19 August 1974

Appendix C. The Outsider view of Nuclear Testing

The following material is from the New Zealand Naval history page at:
http://navymuseum.co.nz/1945-1975-french-nuclear-testing-at-mururoa/

International Intelligence Gathering

The next day, by mistake a United States Navy Sea King helicopter approached the Otago. It hurriedly left and landed aboard the USS Corpus Christi Bay, a helicopter repair ship operated by the Military Sealift command. Along with this there were the RN RFA Sir Percival, USSR research vessels Akademic Shirshov and Volna plus a Chinese fishing vessel gathering signal intelligence![16] All the major nuclear powers had naval forces acting as observers of the test. In this 'great game' of intelligence gathering only the RNZN was acting in a protest role.

Approaching the territorial limits, Otago could see a balloon with the device slung beneath it. They were told to prepare for a test the next day. At 0800 local time, the French detonated a device suspended by balloon at Area Dinton above the atoll at 2000 feet. Otago was 21.5 miles west of the detonation. The flash was intense enough that the crew inside the ship's citadel saw it come through the ventilation system. The yield was estimated to be 5.4 kilotons.[17]. Commander Tyrrell, the navigation officer, the yeoman of signals and two reporters were on the bridge equipped with anti-flash gear and dark goggles.[18] Otago did not detect any radiation with its sensors; it was not affected by any fallout or other contamination from the explosion. Through a radio-telephone link to Wellington, the news was quickly broadcast to the outside world from the reporters aboard Otago. The film footage that was taken would follow in a few days time.

HMNZS Canterbury arrives to Crowded Seas

Canterbury left Auckland to replace Otago on the 14 July she was equipped with the RNZN's first on-board computer nick-named 'Clarence' that monitored the yield of the French bomb and fallout. Despite being hampered by contamination in the port boiler, Canterbury rendezvoused with Otago on 22 July. Otago was ordered back to Mururoa to observe what was thought to be the second test. While Canterbury fixed some engineering issues, Otago remained on duty and moved to a new location for observation. At this point it came across a USN Victory-class intelligence gathering ship, possibly USS Wheeling.[20] The USN ship avoided any contact with the New Zealand frigate. Otago transferred equipment, personnel and Fraser Colman to the Canterbury. Canterbury was then subject to the same level of inspection that Otago had experienced from the surveillance planes.

After a delay noted by the Canterbury from the radio traffic in the morning of 28 July,[21] a device was detonated suspended by balloon 1032 feet at Area Denise. There were some hold-ups in the countdown and an alarm was sounded that caused the French fleet to sail southwards. Canterbury followed in order to avoid the potential fallout zone. The explosion was not heard or seen by men on the Canterbury and was only picked up by the communications team aboard. It was a much smaller yield estimated at .5kt and could not be recorded. Tiny amounts of fallout were recorded and did not pose a danger for the crew. There was some thought that this was a nuclear trigger rather than an operational bomb. The injured master of the detained protest yacht Fri was taken on board on the 3 August. The next morning orders came from Wellington ordering Canterbury home as the release of the yacht Fri was a clear signal that the test programme was concluded for 1973.

Appendix D. Excerpt from Fitness Report

The following excerpt from my fitness report for the period from September 15th, 1973 to the 20th of February 1974 explicitly connects the Claud Jones to Pony Express operations. The report was issued by the Naval Security Group Detachment on the CINCPACFLT staff and was my last report before I was released from active duty. The marks on the front of the form were actually better than I had expected, given the minor disagreement over the performance of the Claud Jones during the last Pony Express operation. I guess the staff didn't want to contradict the letters of appreciation we had received from other commands.

28. DUTIES ASSIGNED (Continued)

NAVSECGRUDET DUTY OFF-3; TEMADD USNS WHEELING {T-AGM-8} 14-24 SEP 73;

TEMADD USS CLAUD JONES {DE-1033} 10-20 DEC 73 AND 31 DEC 73-06 FEB 74.

88. COMMENTS. Particularly comment upon the office's overall leadership ability, personal traits not listed on the reverse side, and estimated or actual performance in combat. Include comments pertaining to unique skills and distinctions that may be important to career development and future assignment. A mark in boxes with an asterisk (*) indicates adversity and supporting comments are required.

LTJG B0RGERS0N IS AN ALERT, INDUSTRIOUS AND CONSCIENTIOUS OFFICER WHO PERFORMED ASSIGNED DUTIES IN AN EXCELLENT MANNER. HE HAS A THOROUGH UNDERSTANDING OF CRYPTOLOGIC SURFACE DIRECT SUPFORT PRODECURES AND RELATED EQUIPMENT CAPABILITIES THAT WAS MANIFESTED IN HIS VALUABLE CONTRIBUTIONS TO THE PLANNING AND EXECUTION OF CRYPTOLOGIC ASPECTS OF SEVERAL FLEET OPERATIONS. HE PERFORMED AS A NAVAL SECURITY GROUP DSE {DIRECT SUPPORT

ELEMENT} DIVISION OFFICER IN USNS WHEELING (TAGM-8) DURING HULA HOOP OPERATIONS CONDUCTED IN JULY-SEPTEMBER 1973. HE SUBSEQUENTLY PARTICIPATED IN A POST-MISSION EVALUATION TO DETERMINE FUTURE NAVAL SECURITY GROUP INVOLVEMENT IN THIS PROGRAM AND FORMULATED COMPREHENSIVE RECOMMENDATIONS TO
IMPROVE THE OPERATIONS. HE DEPLOYED IN USS CLAUD JONES (DE-1033} DURING THE EQUIPMENT TEST PHASE OF OPERATIONS CONDUCTED FOR THE ADVANCED RESEARCH PROJECTS AGENCY AND MADE USEFUL RECOMMENDATIONS CONCERNING THE USE OF PERSONNEL AND RESOURCES. LTJG BORGERSON SUBSEQUENTLY EMBARKED IN CLAUD JONES AS THE DSE DIVISION OFFICER RESPONSIBLE FOR CRYPTOLOGIC SUPPORT TO
PONY EXPRESS OPERATIONS IN THE NORTHERN PACIFIC DURING JANUARY AND FEBRUARY 1974. HIS EXPERIENCE, TECHNICAL AND ADMINISTRATIVE COMPETENCE,AND DEDICATION CONTRIBUTED TO THE OVERALL SUCCESS OF THESE OPERATIONS. HE IS PARTICULARLY THOROUGH IN HIS APPROACH TO ANY PROBLEM AND PERSISTENT IN CARRYING OUT ASSIGNMENTS TO THEIR CONCLUSION. HE HAS DEMONSTRATED THE ABILITY TO ASSUME GREATER RESPONSIBILITIES AND IS RECOMMENDED FOR PROMOTION.

Made in the USA
Middletown, DE
25 March 2021